EXPERIMENTAL AUTOIMMUNE ENCEPHALOMYELITIS – MODELS, DISEASE BIOLOGY AND EXPERIMENTAL THERAPY

Edited by **Robert Weissert**

Experimental Autoimmune Encephalomyelitis – Models, Disease Biology and Experimental Therapy
Edited by Robert Weissert

Published by InTech

Janeza Trdine 9, 51000 Rijeka, Croatia

Publishing Process Manager Oliver Kurelic

Technical Editor Teodora Smiljanic

Cover Designer InTech Design Team

Additional hard copies can be obtained from orders@intechopen.com

Experimental Autoimmune Encephalomyelitis –
Models, Disease Biology and Experimental Therapy, Edited by Robert Weissert
p. cm.
ISBN 978-953-51-0038-6

For EU product safety concerns:
IN TECH d.o.o., Prolaz Marije Krucifikse Kozulić 3, 51000 Rijeka, Croatia,
info@intechopen.com or visit our website at intechopen.com.

Meet the editor

Professor Robert Weissert did his training in Neurology, Neuroimmunology, Immunology and Genetics in Tübingen (Germany), Miami and Chicago (USA) and the Karolinska Institute in Stockholm (Sweden). Between 1999 and 2007, beside his clinical duties as neurologist at the University Hospital in Tübingen, he headed a group of Experimental Neuroimmunology funded by the German Research Foundation (DFG) at the Hertie Institute of Clinical Brain Research in Tübingen. From 2007 to 2010 he worked as a Director in Research and Development of a large pharmaceutical company in Geneva, Switzerland and was a senior physician in the Department of Neurology of the Cantonal Hospital of the University of Geneva. Since 2011 he is working as a senior physician and researcher at the Department of Neurology of the University of Regensburg in Germany. Robert Weissert has made important contributions to the understanding of pathogenesis of immune-mediated diseases and establishment of novel treatment options with focus on multiple sclerosis.

Contents

Preface

`Experimental Autoimmune Encephalomyelitis – Models, Disease Biology and Experimental Therapy` is totally focused on the model of multiple sclerosis, experimental autoimmune encephalomyelitis (EAE). The book chapters give a very good and in depth overview about the currently existing and most used EAE models. In addition, chapters dealing with novel experimental therapeutic approaches demonstrate the usefulness of the EAE model for MS research. MS is a severe disease of the central nervous system (CNS) that leads to progressive neurological deficit. Inflammation in the CNS causes myelin destruction, axonal and neural loss. Even though the disease has been known for centuries, there is only an incomplete understanding of the disease biology, especially of the mechanisms that lead to axonal and neural loss.

With the introduction of interferon-beta preparations in the treatment of MS about 20 years ago, a major therapeutic breakthrough has been achieved in a disease that was considered to be non-treatable. Since then, a number of new treatments have been introduced like Glatirameracetate, Natalizumab and Fingolimod. More are to come. All of these mainly affect inflammatory processes in the CNS. These treatments are not curing the disease but provide much benefit for the patient, reduce relapse rate and severity and slow disease progression. In research and development, a lot of effort is currently being done to discover treatments that are neuroprotective or restoring. Not much light is on the horizon there. EAE is also a very important model that is used in academic research as well as biotech and industry. As outlined in great detail in the different book chapters, different EAE models provide insights into different aspects of MS disease biology and pathology. There is not one model that adequately addresses all facets of MS. For the researcher it is of paramount importance to select the most adequate model for the specific research subject. The book can possibly be of major help in this situation, since, unlike most journal reviews, the book chapters provide a more personal insight into the selection of adequate models. This is an international book with authors contributing from all over the world (Australia, Germany, Japan, Spain, Taiwan, USA). There is an impressive international Faculty that provides insight into current research themes. This further demonstrates the importance of EAE in research all over the world. I am convinced that this book will provide many established researchers and students with novel insights and guidance for their research and will help to push the field forward to better understand the disease biology of MS and other autoimmune diseases and help to establish novel therapeutic approaches.

October 2011

Robert Weissert
University of Regensburg,
Germany

Part 1

Disease Biology

1

Experimental Autoimmune Encephalomyelitis

Robert Weissert
University of Regensburg
Germany

1. Introduction

Experimental autoimmune encephalomyelitis (EAE) is an animal model of multiple sclerosis (MS). MS is also named encephalomyelitis disseminata (ED). EAE has been also called experimental allergic encephalomyelitis, but the word `allergic` is more and more replaced by the word `autoimmune`. This has to do with the deeper understanding of the disease biology that has been gathered over the last years. MS is a chronic autoimmune disease of the central nervous system (CNS) that leads to inflammation, demyelination and axonal loss. The CNS lesions cause neurological deficits.

In the beginning the disease course of MS is in most cases relapsing-remitting but changes after several years into a secondary chronic progressive disease course (Table 1). More rarely, there are also primary progressive disease courses. The diagnosis of MS is often preceded by a single demyelinating event for example the appearance of Optic Neuritis (ON). There are indications that MS is caused by the activation of autoreactive CNS-specific T cells and possibly antibodies that lead to subsequent activation of additional immune cascades and lesion formation (Fig. 1). The relapsing phase of MS is thought to be predominately mediated by the adaptive phase of the immune response, while progressive forms are more driven by innate immune mechanisms (Bhat and Steinman, 2009; Weiner, 2009).

The EAE models that are used in the laboratory for assessment of immunological, neurobiological and therapeutic studies are all induced models or genetically modified models. There is no naturally occurring spontaneous EAE model that is accepted as a valuable laboratory model for MS. EAE can be induced in various animal strains and species. Most used species are mice and rats. The reason for this is the size, the availability of inbred strains and the possibility for genetic modification as well as the immense number of tools to characterize rodents. In addition certain monkey species like marmosets are used for specific questions that cannot be easily assessed in rodents (Hart et al., 2011).

What has become clear over the last years is the fact that EAE is no perfect model for all aspects of MS. Rather various different models represent facets of MS. Some models are more suited for immunological analyses and this can be further divided into adaptive and innate immunity related aspects as well as cellular aspects, like the analysis of influences of T and B cells on disease precipitation and maintenance. There are models that are more suited for analysis of certain neurobiological aspects of the disease, like axonal and neuronal pathology and detailed lesion characterization.

Course	Characteristics	Estimated Prevalence
Relapsing MS (RR-MS, rSP-MS, PR-MS)		
RR-MS (Relapsing-remitting MS)	• Presence of relapses and remissions • Disability due to residual symptoms	22 %
rSP-MS (SP-MS with superimposed relapses)	• Presence of relapses • Progressive form, steadily increasing disability	3.5 %
PR-MS (Progressive relapsing MS)	• Presence of relapses • Progressive from onset, steadily increasing disability	1.5 %
Progressive MS without relapses (nrSP-MS, PP-MS)		
nrSP-MS (SP-MS without superimposed relapses)	• No relapses • Progressive form, steadily increasing disability	60 %
PP-MS (Primary progressive MS)	• No relapses • Progressive from onset, steadily increasing disability	13 %

Table 1. Different disease courses of MS. Relapsing disease courses of MS are mainly driven by the adaptive immune response (T and B cells) while progressive disease is thought to be predominantly mediated by innate immunity. In most cases, relapsing disease changes over the course of MS to secondary progressive disease. As outlined in Table 10, specific EAE models can be used to mimic the aspects of different types and variants (Devic´s, ADEM) of MS.

MS variants can be modeled in rodents. For example Acute Disseminated Encephalomyelitis (ADEM), an acute inflammatory reactive disease of the CNS, and Neuromyelitis Optica (Devic`s disease) can be modeled in certain rodent species as outlined further below. The researcher who wants to use EAE as a model should be aware of all the various types of EAE and should be careful in selecting the correct species, strain and immunization protocol, since depending on these factors, the outcome and interpretability of the research will differ. The selection of the specific model might differ strongly between an immunologist who wants to assess basic principles of organ specific autoimmunity in contrast to the MS researcher who wants to analyze specific disease aspects or a new therapeutic approach in preclinical pharmacology studies.

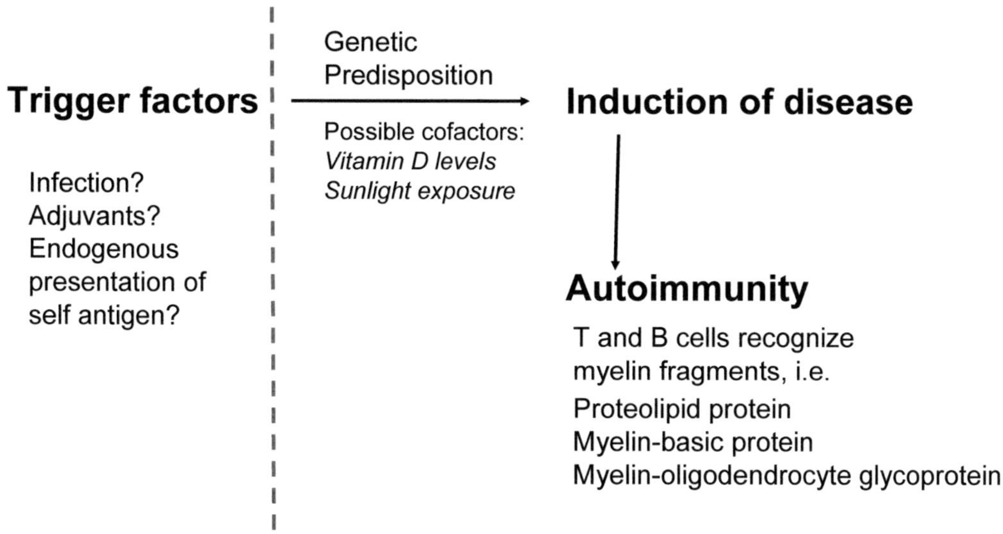

Fig. 1. Possible cascade for induction of MS. So far it is not clear what induces MS. Trigger factors are possibly infections, adjuvants or the endogenous presentation of self antigens (Fissolo et al., 2009). On the ground of a genetic predisposition with the possible cofactors vitamin D and sunlight exposure the disease is induced and autoimmunity is maintained by aberrant immune cascades that will finally lead to myelin/oligodendrocyte pathology as well as axonal and neuronal loss.

2. EAE models

Nowadays widely used EAE models in rats are the monophasic EAE in LEW rats that is induced with Myelin Basic Protein (MBP) or MBP peptides and Complete Freund´s Adjuvant (CFA) (Table 2). CFA consists of the mineral oil Incomplete Freund´s Adjuvant (IFA) mixed with heat killed extract from Mycobacterium Tuberculosis (MT). As prototype chronic EAE model in rats, EAE induced in DA rats with whole myelin extract in IFA or the extracellular domain of Myelin Oligodendrocyte Glycoprotein (MOG) in CFA or IFA have high value for studies regarding MS disease biology and therapeutic interventions. Whole myelin extract for this model is typically obtained from autologous spinal cord of DA rats. Interestingly DA rats develop EAE with the myelin extract/MOG with IFA alone.

In mice the most used model is the chronic model in C57BL/6 mice induced with the extracellular domain of MOG or MOG peptide 35-55 and CFA as well as Pertussis Toxin (PT) as adjuvant (Table 2). Other valuable models are the chronic relapsing EAE model induced with PLP or PLP peptide 139-151 in SJL mice with CFA and PT as ajduvants, the chronic relapsing EAE model induced in Biozzi mice with the extracellular domain of MOG or MOG peptide 92-106 in CFA and PT, the chronic progressive EAE induced in autoimmunity prone NOD mice with MOG peptide 35-55 and the monophasic EAE induced with MBP peptide Ac1-11 (or MBP peptide Ac1-9) with CFA and PT in PL/J mice.

There are relapsing progressive EAE models that can be induced with Theiler`s Murine Encephalitis Virus (TMEV) (Table 2). Interestingly in the TMEV model, antigen spreading to the classical myelin antigens can be observed during the disease course. Genetically

Species	Strain	Induction	Disease type	Reference
Rat	LEW	MBP or MBP 68-88 peptide in CFA	Acute monophasic	(Mannie et al., 1985; McFarlin et al., 1975)
Rat	DA	Spinal cord in IFA or MOG 1-125 in CFA or IFA	Relapsing/progressive	(Lorentzen et al., 1995; Weissert et al., 1998b)
Mouse	PL/J	MBP, MBP Ac1-11 (or Ac1-9) in CFA + PT	Acute monophasic	(Zamvil et al., 1986)
Mouse	SJL	PLP protein or PLP 139-151 peptide in CFA + PT	Chronic relapsing	(Kuchroo et al., 1991)
Mouse	C57BL/6	MOG protein or MOG 35-55 peptide in CFA + PT	Progressive or acute	(Mendel et al., 1995; Oliver et al., 2003)
Mouse	Biozzi	MOG protein or MOG 92-106 peptide in CFA + PT	Chronic relapsing	(Baker et al., 1990)
Mouse	NOD	MOG 35-55 peptide in CFA + PT	Chronic progressive	(Maron et al., 1999)
Mouse	SJL	Theiler's murine encephalitis virus	Relapsing progressive	(Olson et al., 2001)
Mouse	B10.PL-$H2^{u}$	TCR^{MBP} transgenic	Spontaneous acute	(Goverman et al., 1993)
Mouse	C57BL/6	TCR^{MOG} transgenic	Optic neuritis	(Bettelli et al., 2003)
Mouse	C57BL/6	$TCR^{MOG}XIgH^{MOG}$ transgenic	Spontaneous Optico-spinal disease	(Bettelli et al., 2006; Krishnamoorthy et al., 2006)

Table 2. Presently most used EAE models for laboratory research. The list only provides a part of all available EAE models. CFA = Complete Freund`s Adjuvant; IFA = Incomplete Freund`s Adjuvant; PT = Pertussis Toxin.

modified models include the spontaneous EAE in C57BL/6 mice that have a T Cell Receptor (TCR) specific for MBP peptide 1-9. Another genetically modified C57BL/6 mouse with a MOG specific TCR and a MOG specific B Cell Receptor (BCR) or surface Immunoglobulin (sIg) develop spontaneous EAE and can be very useful for assessment of specific scientific questions.

There are rodent EAE models that can be induced by passive transfer of T cells. These models are not subject of this chapter and are therefore not outlined in detail. T cell transfer models have been very suitable for immunological investigations mainly in regard to the dissection of the role of T cells in organ-specific autoimmunity (Krishnamoorthy and Wekerle, 2009).

3. Establishment of Myelin-Oligodendrocyte-Glycoprotein (MOG)- induced EAE in rats

The rat is a good species to perform EAE studies due to its larger size as compared to mice, the availability of many well characterized inbred strains, its strong standing for pharmacological studies and its great value for behavioral outreads. In addition novel ways for genetic modification are increasingly becoming available. The rat Major Histocompatibility Complex (MHC) is called RT1. The classical MHC I molecule is called RT1.A, the MHC II molecules are named RT1.B and RT1.D that are equivalent to HLA-DQ (RT1.B) and HLA-DR (RT1.D) (Fig. 2, Table 3).

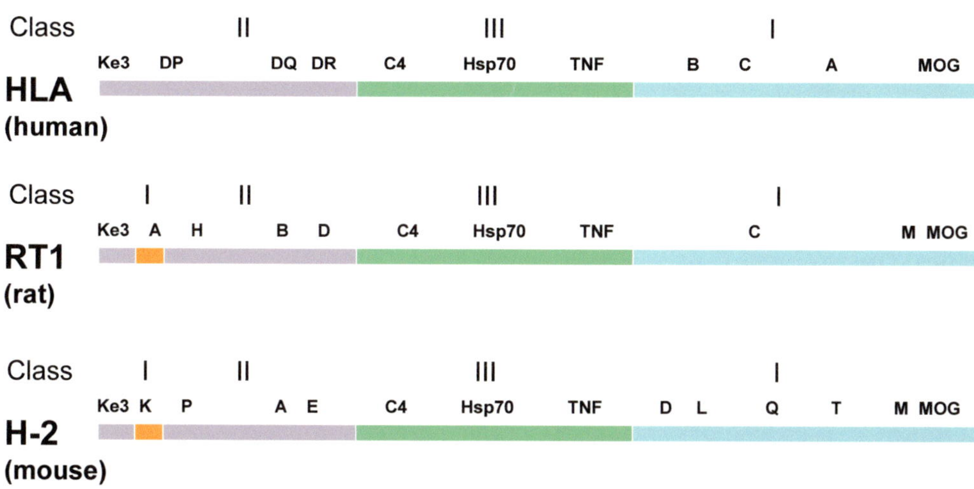

Fig. 2. Organization of MHC in different species (human [HLA], rat [RT1], mice [H2]). There are considerable differences in the organization between HLA, H2 and RT1 (Günther and Walter, 2001).

EAE induced with whole myelin extracts or MBP in LEW ($RT1^l$) rats has been used over many years (Table 2). This model is monophasic and was the prototype EAE model in the past. Much understanding regarding MS pathogenesis, immunology and neurobiology has been obtained over the years. The model is very useful to study basic immunological principles regarding T cell migration to the CNS and T cell related immunity of MS. This model is often used for pharmacological investigations in EAE. Beside with whole myelin extracts the model can be induced with MBP protein (various isoforms) or peptides from the main encephalitogenic regions MBP 89-101 and MBP 68-88. Interestingly MBP peptide 89-101 binds to $RT1.D^l$ (HLA-DR-like), while MBP 68-88 binds to $RT1.B^l$ (HLA-DQ-like) (de Graaf et al., 1999; Weissert et al., 1998a). There are species dependent effects of MBP on disease development. We could demonstrate that the MBP peptide 63-88 derived from Guinea Pig (GP) as compared to the MBP peptide 63-88 derived from rat is much more encephalitogenic. Immunization with MBP_{GP}63-88 results in the agglomeration of T cells in the CNS that express up to 30% the TCRBV chain 8.2 (TCRBV8S2), while after immunization with MBP_{RAT}63-88 there is not such a strong overusage of such T cells in the CNS (Weissert et al., 1998a). Based on this fact it can also be understood that therapeutic manipulation in the context of immunization with MBP_{GP}63-88 is more demanding as compared to MBP_{RAT}63-88. We equally demonstrated that MBP derived peptides of different species can act as superagonists, agonists or antagonists depending on the expressed MHC II haplotype (de Graaf et al., 2005).

Strain*		*Class I*	*Class II*		*Class III*	*Class I*
	Haplotyp	RT1.A	RT1.B	RT1.D		RT1.C
DA	av1	a	a	a	av1	av1
COP	av1	a	a	a	av1	av1
ACI	av1	a	a	a	av1	av1
PVG-$RT1^a$ (DA)	av1	a	a	a	av1	av1
LEW.1AV1 (DA)	av1	a	a	a	av1	av1
LEW	l	l	l	l	l	l
LEW.1N (BN)	n	n	n	n	n	n
LEW.1A (AVN)	a	a	a	a	a	a
LEW.1W (WP)	u	u	u	u	u	u
LEW.1AR1	r2	a	u	u	u	u
LEW.1AR2	r3	a	a	a	u	u
LEW.1WR1	r4	u	u	u	a	a
LEW.1WR2	r6	u	a	a	a	a
BN	n	n	n	n	n	n

Table 3. RT1 haplotypes of inbred rat strains (Weissert et al., 1998b). RT1.B is the rat equivalent to HLA-DQ and RT1.D to HLA-DR. *Donor strain in brackets

It was demonstrated that as compared to LEW rats, DA rats develop in a much higher incidence chronic disease after immunization with autologous whole spinal cord homogenates (Lorentzen et al., 1995). In DA rats immunization of autologous whole spinal cord in IFA is sufficient to induce this type of chronic disease. Interestingly it was found that DA rats develop antibodies against the MOG extracellular domain MOG 1-125. Based on this fact we were interested to assess the encephalitogenic potential of MOG 1-125 in DA rats and other inbred rat strains. We found that immunization with MOG 1-125 in CFA resulted in relapsing-remitting disease (Weissert et al., 1998b). Also the immunization of MOG 1-125 in IFA causes to this type of disease. Interestingly for the first time it was observed in rodents that the rats developed ON in addition to classical EAE symptoms (Storch et al., 1998). In addition, specific immunization protocols allowed the selective induction of ON. ON is a typical aspect of MS. The CNS lesions in this model have much greater similarity to MS as compared to the lesions in LEW rats immunized with MBP or MBP peptides. Widespread demyelination was present and the analysis of lesions resulted in the important insight that beside T cells also antibodies contribute to lesion formation in this model. In contrast to MBP, MOG is expressed on the exterior part of the myelin sheath and the Ig-like domain can be recognized by antibodies. Binding of antibodies to MOG can activate a number of immune mechanisms like the activation of macrophages and complement deposition that can result in increased tissue damage as compared to purely T cell mediated pathology.

Based on the findings in DA rats and immunization with MOG we were wondering about the encephalitogenic potential in rat strains with different MHC haplotypes. This question appeared important to us, since it has been demonstrated for a long time that the MHC II region has a strong genetic influence on MS (Sawcer et al., 2011). Therefore we used a large number of inbred congenic LEW rat strains that express a wide variety of MHC haplotypes. In addition we used various inbred rat strains (Table 3) (Weissert et al., 1998b).

We observed that certain rat strains developed relapsing-remitting or chronic progressive disease, others hyperacute progressive disease, some slow progressive disease with predominance of cortical pathology and others were protected (Table 4). Also the selection of the adjuvant had additional effects. While LEW.1N ($RT1^n$) immunized with MOG 1-125 in CFA developed hyperacute progressive disease that often lead to death due to pontine lesions, immunization with MOG 1-125 in IFA resulted in chronic progressive disease with axonal pathology that has great similarity to MS (Kornek et al., 2000).

We also observed that LEW ($RT1^l$) rats immunized with MOG 1-125 did not develop EAE (Weissert et al., 1998b). This is in contrast to immunization protocols with MBP or MBP peptides which lead to EAE as outlined above (Weissert et al., 1998a). Based on the fact that LEW.1AV1 rats that carry the $RT1^{av1}$ haplotype derived from the DA rat develop EAE after immunization with MOG 1-125, we concluded that the MHC haplotype is operating in the context of the myelin antigen used for immunization (Weissert et al., 1998b). This was further supported by the observation that LEW.1N ($RT1^n$) rats and BN ($RT1^n$) rats do not develop disease after immunization with MBP, but strong disease after immunization with MOG 1-125. This finding appears of some importance since it might explain why MHC haplotypes might differ as susceptibility loci in different parts of the world: depending on the environmental challenges that differ in different regions of the world, the subsequent induction of immunity against certain myelin components might differ. For example in Western Europe and North America the main susceptibility HLA allele is HLA-DR2b, while in Sardinia it is HLA-DR4 (Marrosu et al., 1997).

We described rat strains that carried MHC haplotypes that allow disease development after immunization with MOG 1-125 but which were protected by non-MHC genomes. We defined that ACI rats as well as MHC congenic PVG-RT1^{av1} that both carry the RT1^{av1} haplotpye were protected from EAE (Weissert et al., 1998b). Subsequent large genetic screens allowed dissection of some of the susceptibility genes (Swanberg et al., 2005).

Species	Strain	Induction	Disease type	Reference
Rat	LEW.1A	MOG1-125 in CFA	Chronic progressive	(Weissert et al., 1998b)
Rat	LEW.1AV1	MOG1-125 in CFA	Relapsing-remitting	(Weissert et al., 1998b)
Rat	LEW.1AV1	MOG1-125 in IFA	Relapsing-remitting	(Kornek et al., 2000)
Rat	LEW.1N	MOG1-125 in CFA	Hyperacute progressive	(Weissert et al., 1998b)
Rat	LEW.1N	MOG1-125 in IFA	Chronic progressive	(Kornek et al., 2000)
Rat	LEW.1W	MOG1-125 in CFA	Slow progressive	(Weissert et al., 1998b)
Rat	LEW.1AR1	MOG1-125 in CFA	Slow progressive, cortical pathology	(Storch et al., 2006; Weissert et al., 1998b)
Rat	DA	MOG1-125 in CFA	Relapsing-remitting, optic neuritis	(Storch et al., 1998; Weissert et al., 1998b)
Rat	BN	MOG1-125 in CFA	Neuromyelitis optica	(Meyer et al., 2001; Weissert et al., 1998b)

Table 4. EAE models induced with MOG 1-125 in rats. Different disease courses and types of CNS pathology can be induced that are dependent on the MHC haplotype.

We found one LEW rat strain with a specific MHC haplotype that predominantly develops cortical lesions, LEW.1AR1 (RT1^{r2}) rats (Storch et al., 2006; Weissert et al., 1998b). Recently cortical lesions have been acknowledged as a primary cause of disability in MS. Cortical lesions can be also induced in LEW rats with a subencephalitogenic immunization with myelin components and subsequent local intrathecal application of Tumor Necrosis Factor alpha (TNFα) (Merkler et al., 2006). Marmosets immunized with MOG 1-125 can develop cortical lesions as well (Pomeroy et al., 2005). By now it has also been observed that certain mouse strains can develop cortical pathology (Mangiardi et al., 2011). In our observation LEW.1AR1 rats immunized with MOG 1-125 in CFA represent the most suitable and reproducible model for cortical pathology of MS and possible therapeutic manipulation.

In a next step we were interested to define the encephalitogenic stretches within the extracellular domain of MOG 1-125 (Tables 5, 6). In order to do this we used 18 amino acid long overlapping peptides of the MOG 1-125 rat sequence (Weissert et al., 2001). We performed this study in many different inbred and inbred MHC congenic rat strains. Interestingly we found that only MOG 91-108 in CFA induced disease in rat strains with the $RT1^{av1}$ or $RT1^{n}$ haplotype. Rat strains with the $RT1^{av1}$ haplotype are DA rats and LEW.1AV1 rats and rats with the $RT1^{n}$ haplotype LEW.1N and BN rats. We found that MOG 91-108 bound well to purified $RT1.B^{a}$ molecules and to $RT1.D^{n}$ molecules. Immunization with MOG 91-108 induced a T and B cell response. In LEW.1N rats this T cell response was difficult to measure and we concluded early that other factors than classical Th1 mediated cytokines might be operative contributing to disease precipitation (Weissert et al., 2001).

In most instances the immunodominant peptides did not correspond to the peptides that were capable of inducing EAE. We concluded that the immune response to the immunodominant peptides might be also a signature of a regulatory T cell response (we called it `modulatory).

RT1	RT1.A	RT1.B/D	RT1.C	Strain	Disease inducing peptides	Peptides that raise an immuno-dominant T cell response
1	1	1	1	LEW	none	MOG 37-54 and MOG 43-60
u	u	u	u	LEW.1W	none	None
r4	u	u	a	LEW.1WR1		
r2	a	u	u	LEW.1AR1		
a	a	a	a	LEW.1A	MOG 91-108, MOG 96-104	MOG 73-90, MOG 91-108
r3	a	a	u	LEW.1AR2		
r6	u	a	a	LEW.1WR2		
av1	a	a	av1	LEW.1AV1		
av1	a	a	av1	DA		
av1	a	a	av1	COP		
av1	a	a	av1	PVT-$RT1^{av1}$	none	none
av1	a	a	av1	ACI		
n	n	n	n	LEW.1N	MOG 91-108, MOG 98-106	MOG 19-36
n	n	n	n	BN		

Table 5. MOG peptides that induce disease and immune responses. Most MOG stretches that induce strong immune responses in rats do not induce EAE (Weissert et al., 2001).

Species	Strain	Induction	Disease type	Reference
Rat	DA	MOG91-108 in CFA	Monophasic or chronic	(Weissert et al., 2001)
Rat	LEW.1AV1	MOG91-108 in CFA	Monophasic or chronic	(Weissert et al., 2001)
Rat	LEW.1N	MOG91-108 in CFA	Monophasic or chronic	(Weissert et al., 2001)
Rat	LEW.1AV1	MOG96-104 in CFA	Monophasic or chronic	(de Graaf et al., 2008)
Rat	LEW.1N	MOG98-106 in CFA	Monophasic or chronic	(de Graaf et al., 2008)

Table 6. EAE models induced with MOG peptides. Based on detailed immunological analysis the region MOG 91-108 was defined as the encepathalitogenic region in different rat strains. Further dissection allowed the narrowing of the disease inducing MOG stretches to nine amino acid long peptides.

We observed that complement depletion does lead to protection from disease, underscoring the influence of antibodies to MOG. Crystallographic studies have indicated that the region MOG 91-108 is accessible to antibodies (Breithaupt et al., 2008). In addition, we demonstrated by spectroscopic TCR analysis that depending on the expressed RT1 haplotype, the predominance of certain TCRBV chains was the same in rat strains with different non-MHC genomes underscoring the strong influence of the MHC II haplotype on TCRBV usage in the MOG model (de Graaf et al., 2008).

For the rat EAE models, we measured the binding strength of the encephalitogenic peptides by comptetitive binding assays (Table 7). We could show that the peptides that induce EAE are binding well to the MHC II molecules that present the peptides to T cells (de Graaf et al., 2008; de Graaf et al., 1999; Weissert et al., 2001; Weissert et al., 1998a). We dissected the binding qualities to purified RT1.B and RT1.D molecules and the T and B cell response to MOG 91-108 in LEW.1N and LEW.1AV1 rats. We found that the peptides that bound strongest, induced EAE. This were the peptides MOG 96-104 in LEW.1AV1 rats binding to RT1.B^a and MOG 98-106 in LEW.1N rats binding to RT1.D^n. With increasing shortening of the peptides, the evolving disease was partly reduced, indicating that possibly there was a reduction in the activated encephalitogenic T cell repertoire (de Graaf et al., 2008).

That peptides which induce EAE bind strongly to the restricting MHC II molecule is in agreement with measured binding strengths in humanized mouse models, but contrasts findings in the PL/J mouse in which the encephalitogenic peptide MBP Ac1-9 binds only very weakly to the I-A^u MHC II molecule. While in the first case the persistence of antigen might lead to breaking of tolerance, in the latter case, the escape from tolerance might be of primary importance in disease establishment.

Haplotype (Strain)	Encephalitogen	Restriction	IC_{50} (µM)	Reference
$RT1^{l}$ (LEW)	MBP_{gp}72-85	B^{l}	2,5	(Weissert et al., 1998a)
	MBP 87-99	D^{l}	0,02	(de Graaf et al., 1999)
$RT1^{av1}$ (DA, LEW.1AV1)	MBP 87-99	B^{a}	0,06	(de Graaf et al., 1999)
	MOG 91-108, MOG 96-104	B^{a}	9,5	(de Graaf et al., 2008; Weissert et al., 2001)
$RT1^{n}$ (BN, LEW.1N)	MOG 91-108, MOG 98-106	D^{n}	0,03	(de Graaf et al., 2008; Weissert et al., 2001)
$H\text{-}2^{s}$ (SJL/J)	MBP 81-100	$I\text{-}A^{s}$	0,36	(Wall et al., 1992)
	PLP 100-119	$I\text{-}A^{s}$	1,24	(Greer et al., 1996)
	PLP 139-151	$I\text{-}A^{s}$	0,04	(Greer et al., 1996)
	PLP 178-191	$I\text{-}A^{s}$	0,74	(Greer et al., 1996)
$H\text{-}2^{u}$ (PL/J)	MBP Ac1-9	$I\text{-}A^{u}$	> 100	(Fairchild et al., 1993; Liu et al., 1995)
HLA-DR2 transgenic mice	MBP 84-102	DRB1*1501	0,004	(Madsen et al., 1999; Wucherpfennig et al., 1994)
HLA-DR4 transgenic mice	MOG 91-108	DRB1*0401	0,3	(Forsthuber et al., 2001)

Table 7. Binding of EAE inducing peptides to purified MHC molecules. Binding strengths were assessed with competitive binding assays and affinity purified MHC II molecules.

In summary we made a major step in establishing more suitable models to study MS disease biology and therapeutic interventions. In addition we were able to obtain a large quantity of insight into the immune regulation in the context of genetic factors in a complex autoimmune disease. We demonstrated the major influence of MHC II haplotypes on disease regulation, but for the first time also of MHC I. Recently also for MS influences of MHC I loci on susceptibility could be confirmed (Sawcer et al., 2011).

4. Ways of EAE induction with either MBP, PLP or MOG and choice of the relevant animal model/strain

Beside the selection of the right species and strain, the selection of the model antigen is of great importance for the outcome of the EAE studies. The sequences of the most used stretches of myelin proteins for disease induction used in different species are listed in Table 8. The peptides should have a high degree of purity and it is advisable to prepare large batches that can be used over long term for experimentation. In the case of EAE induction with recombinant antigens, also larger scale preparation and usage of identical batches over longer time is recommendable, since there is the danger of batch to batch variation that can dramatically affect the outcome of experimentation.

In addition to species, strain and antigen, the adjuvant is of major importance for the success of the EAE induction and the outcome of the experimentation (Table 9). While in the rat models, the usage of IFA or CFA is sufficient, nearly all mouse models require the addition of PT in the immunization protocol and booster immunizations. The preparation of the antigen/adjuvant mixture requires much care and the presence of a homogenate emulsion is needed for successful immunizations.

In the past immunizations have often been performed in the foot pads of the rodents. Due to obvious ethical reasons, this procedure is not any more applied. In addition this procedure results in the swelling of the footpads affecting gait. This type of gait disturbance can blur EAE symptoms with the consequence of wrongly reported EAE scores. Nowadays, in rats the immunization is done as a single injection in the base of the tail, while in mice multiple injections in the flanks are used for the procedure. It is advisable to establish the best suited immunization protocol in the laboratory with care, after the selection of the model based on scientific rationales has been performed.

Myelin protein stretch	**Sequence**
MBP_{MOUSE} Ac1-9	Ac-ASQKRPSQR
PLP_{MOUSE} 139-151	HSLGKWLGHPDK
MOG_{MOUSE} 35-55	MEVGWYRSPFSRVVHLYRNGK
MBP_{RAT} 68-88	HYGSLPQKSQRTQDENPVVHF
MBP_{GP}68-88	HYGSLPQKSQRSQDENPVVHF
MBP_{RAT} 89-101	VHFFKNIVTPRTP
MOG_{RAT} 91-108	SDEGGYTCFFRDHSYQEE
MOG_{RAT} Ac96-104	Ac-YTCFFRDHS-NH2
MOG_{RAT}Ac98-106	Ac-CFFRDHSYQ-NH2

Table 8. Sequences of myelin peptides used for EAE induction in mice and rats

Species/ strain	Immunogen	Primary adjuvant	Secondary adjuvant	Injection site	Reference
Mouse					
PL/J	MBP Ac1-11	CFA	PT	Flanks	(Zamvil et al., 1986)
SJL	PLP 139-151	CFA	PT	Flanks	(Kuchroo et al., 1991)
Biozzi	MOG 1-125	CFA	PT	Flanks	(Baker et al., 1990)
C57Bl/6	MOG 35-55	CFA	PT	Flanks	(Oliver et al., 2003)
Rat					
LEW	MBP	CFA	None	Tail base	(Weissert et al., 2000)
LEW	MBP 68-88	CFA	None	Tail base	(Weissert et al., 1998b)
LEW	MBP 89-101	CFA	None	Tail base	(Weissert et al., 2000)
LEW	PLP	CFA	None	Tail base	(Zhao et al., 1994)
DA, LEW.1AV1, BN, LEW.1N, LEW1.AR1	MOG 1-125	CFA or IFA	None	Tail base	(Kornek et al., 2000; Storch et al., 2006; Weissert et al., 1998b)
DA, LEW.1AV1	MOG 91-108, MOG 96-104	CFA	None	Tail base	(de Graaf et al., 2008; Weissert et al., 2001)
LEW.1N or BN	MOG 91-108, MOG 98-106	CFA	None	Tail base	(de Graaf et al., 2008; Weissert et al., 2001)

Table 9. Immunization regimen for different EAE models and usage of adjuvants. CFA = Complete Freund`s Adjuvant; IFA = Incomplete Freund`s Adjuvant; PT = Pertussis Toxin.

5. Conclusions

It is well possible to induce different aspects of MS in rodent EAE models (Table 10). Some models are more suited for immunological analysis, while others better serve the neuroscience community. None of the models can model all aspects of MS. Based on the specific scientific question, the most suitable EAE model for a specific analysis should be selected based on the characteristics of the model. The selection of the best suited model will result in better results of the overall research project and will improve the interpretability of the results.

Course of MS	Type of EAE	Strain	Immunization
Relapsing forms of MS			
RR-MS	MOG-EAE	DA, LEW.1AV1	MOG 1-125
rSP-MS	MOG-EAE	DA	MOG 1-125
PR-MS	MOG-EAE	LEW.1N	MOG 1-125
Non-relapsing forms of MS			
nrSP-MS	MOG-EAE	LEW.1A	MOG 1-125
PP-MS	MOG-EAE	LEW.1W, LEW.1WR1	MOG 1-125
Specific pathologies or rare MS variants			
Predominance of cortical lesions	MOG-EAE	LEW.1AR1	MOG 1-125
Devic`s disease	MOG-EAE	BN	MOG 1-125
ADEM	MBP-EAE	LEW	MBP_{GP}68-88, MBP_{RAT}68-88

Table 10. Best suited rat model to investigate aspects of different MS types and MS variants

The MOG-EAE model in rats with the availability of various inbred and RT1 congenic strains provides a very good and well defined system for assessment of pertinent questions regarding MS disease biology and therapeutic interventions.

6. References

Baker, D., O'Neill, J. K., Gschmeissner, S. E., Wilcox, C. E., Butter, C., and Turk, J. L. (1990). Induction of chronic relapsing experimental allergic encephalomyelitis in Biozzi mice. J Neuroimmunol *28*, 261-270.

Bettelli, E., Baeten, D., Jager, A., Sobel, R. A., and Kuchroo, V. K. (2006). Myelin oligodendrocyte glycoprotein-specific T and B cells cooperate to induce a Devic-like disease in mice. J Clin Invest *116*, 2393-2402.

Bettelli, E., Pagany, M., Weiner, H. L., Linington, C., Sobel, R. A., and Kuchroo, V. K. (2003). Myelin oligodendrocyte glycoprotein-specific T cell receptor transgenic mice develop spontaneous autoimmune optic neuritis. J Exp Med *197*, 1073-1081.

Bhat, R., and Steinman, L. (2009). Innate and adaptive autoimmunity directed to the central nervous system. Neuron *64*, 123-132.

Breithaupt, C., Schafer, B., Pellkofer, H., Huber, R., Linington, C., and Jacob, U. (2008). Demyelinating myelin oligodendrocyte glycoprotein-specific autoantibody response is focused on one dominant conformational epitope region in rodents. J Immunol *181*, 1255-1263.

de Graaf, K. L., Barth, S., Herrmann, M. M., Storch, M. K., Wiesmuller, K. H., and Weissert, R. (2008). Characterization of the encephalitogenic immune response in a model of multiple sclerosis. Eur J Immunol *38*, 299-308.

de Graaf, K. L., Berne, G. P., Herrmann, M. M., Hansson, G. K., Olsson, T., and Weissert, R. (2005). CDR3 sequence preference of TCRBV8S2+ T cells within the CNS does not reflect single amino acid dependent avidity expansion. J Neuroimmunol *166*, 47-54.

de Graaf, K. L., Weissert, R., Kjellen, P., Holmdahl, R., and Olsson, T. (1999). Allelic variations in rat MHC class II binding of myelin basic protein peptides correlate with encephalitogenicity. Int Immunol *11*, 1981-1988.

Fairchild, P. J., Wildgoose, R., Atherton, E., Webb, S., and Wraith, D. C. (1993). An autoantigenic T cell epitope forms unstable complexes with class II MHC: a novel route for escape from tolerance induction. Int Immunol *5*, 1151-1158.

Fissolo, N., Haag, S., de Graaf, K. L., Drews, O., Stevanovic, S., Rammensee, H. G., and Weissert, R. (2009). Naturally presented peptides on major histocompatibility complex I and II molecules eluted from central nervous system of multiple sclerosis patients. Mol Cell Proteomics *8*, 2090-2101.

Goverman, J., Woods, A., Larson, L., Weiner, L. P., Hood, L., and Zaller, D. M. (1993). Transgenic mice that express a myelin basic protein-specific T cell receptor develop spontaneous autoimmunity. Cell *72*, 551-560.

Greer, J. M., Sobel, R. A., Sette, A., Southwood, S., Lees, M. B., and Kuchroo, V. K. (1996). Immunogenic and encephalitogenic epitope clusters of myelin proteolipid protein. J Immunol *156*, 371-379.

Günther, E., and Walter, L. (2001). The major histocompatibility complex of the rat (Rattus norvegicus). Immunogenetics *53*, 520-542.

Hart, B. A., Gran, B., and Weissert, R. (2011). EAE: imperfect but useful models of multiple sclerosis. Trends Mol Med. *17*, 119-125.

Kornek, B., Storch, M. K., Weissert, R., Wallstroem, E., Stefferl, A., Olsson, T., Linington, C., Schmidbauer, M., and Lassmann, H. (2000). Multiple sclerosis and chronic autoimmune encephalomyelitis: a comparative quantitative study of axonal injury in active, inactive, and remyelinated lesions. Am J Pathol *157*, 267-276.

Krishnamoorthy, G., Lassmann, H., Wekerle, H., and Holz, A. (2006). Spontaneous opticospinal encephalomyelitis in a double-transgenic mouse model of autoimmune T cell/B cell cooperation. J Clin Invest *116*, 2385-2392.

Krishnamoorthy, G., and Wekerle, H. (2009). EAE: an immunologist's magic eye. Eur J Immunol *39*, 2031-2035.

Kuchroo, V. K., Sobel, R. A., Yamamura, T., Greenfield, E., Dorf, M. E., and Lees, M. B. (1991). Induction of experimental allergic encephalomyelitis by myelin proteolipid-protein-specific T cell clones and synthetic peptides. Pathobiology *59*, 305-312.

Liu, G. Y., Fairchild, P. J., Smith, R. M., Prowle, J. R., Kioussis, D., and Wraith, D. C. (1995). Low avidity recognition of self-antigen by T cells permits escape from central tolerance. Immunity *3*, 407-415.

Lorentzen, J. C., Issazadeh, S., Storch, M., Mustafa, M. I., Lassman, H., Linington, C., Klareskog, L., and Olsson, T. (1995). Protracted, relapsing and demyelinating experimental autoimmune encephalomyelitis in DA rats immunized with syngeneic spinal cord and incomplete Freund's adjuvant. J Neuroimmunol *63*, 193-205.

Madsen, L. S., Andersson, E. C., Jansson, L., krogsgaard, M., Andersen, C. B., Engberg, J., Strominger, J. L., Svejgaard, A., Hjorth, J. P., Holmdahl, R., *et al.* (1999). A humanized model for multiple sclerosis using HLA-DR2 and a human T-cell receptor. Nat Genet *23*, 343-347.

Mangiardi, M., Crawford, D. K., Xia, X., Du, S., Simon-Freeman, R., Voskuhl, R. R., and Tiwari-Woodruff, S. K. (2011). An animal model of cortical and callosal pathology in multiple sclerosis. Brain Pathol *21*, 263-278.

Mannie, M. D., Paterson, P. Y., U'Prichard, D. C., and Flouret, G. (1985). Induction of experimental allergic encephalomyelitis in Lewis rats with purified synthetic peptides: delineation of antigenic determinants for encephalitogenicity, in vitro activation of cellular transfer, and proliferation of lymphocytes. Proc Natl Acad Sci U S A *82*, 5515-5519.

Maron, R., Hancock, W. W., Slavin, A., Hattori, M., Kuchroo, V., and Weiner, H. L. (1999). Genetic susceptibility or resistance to autoimmune encephalomyelitis in MHC congenic mice is associated with differential production of pro- and anti-inflammatory cytokines [In Process Citation]. Int Immunol *11*, 1573-1580.

Marrosu, M. G., Murru, M. R., Costa, G., Cucca, F., Sotgiu, S., Rosati, G., and Muntoni, F. (1997). Multiple sclerosis in Sardinia is associated and in linkage disequilibrium with HLA-DR3 and -DR4 alleles. Am J Hum Genet *61*, 454-457.

McFarlin, D. E., Hsu, S. C., Slemenda, S. B., Chou, F. C., and Kibler, R. F. (1975). The immune response against myelin basic protein in two strains of rat with different genetic capacity to develop experimental allergic encephalomyelitis. J Exp Med *141*, 72-81.

Mendel, I, Kerlero de Rosbo, N., and Ben-Nun, A. (1995). A myelin oligodendrocyte glycoprotein peptide induces typical chronic experimental autoimmune encephalomyelitis in H-2b mice: fine specificity and T cell receptor V beta expression of encephalitogenic T cells. Eur J Immunol *25*, 1951-1959.

Merkler, D., Ernsting, T., Kerschensteiner, M., Bruck, W., and Stadelmann, C. (2006). A new focal EAE model of cortical demyelination: multiple sclerosis-like lesions with rapid resolution of inflammation and extensive remyelination. Brain *129*, 1972-1983.

Meyer, R., Weissert, R., Diem, R., Storch, M. K., de Graaf, K. L., Kramer, B., and Bahr, M. (2001). Acute neuronal apoptosis in a rat model of multiple sclerosis. J Neurosci *21*, 6214-6220.

Oliver, A. R., Lyon, G. M., and Ruddle, N. H. (2003). Rat and human myelin oligodendrocyte glycoproteins induce experimental autoimmune encephalomyelitis by different mechanisms in C57BL/6 mice. J Immunol *171*, 462-468.

Olson, J. K., Croxford, J. L., Calenoff, M. A., Dal Canto, M. C., and Miller, S. D. (2001). A virus-induced molecular mimicry model of multiple sclerosis. J Clin Invest *108*, 311-318.

Pomeroy, I. M., Matthews, P. M., Frank, J. A., Jordan, E. K., and Esiri, M. M. (2005). Demyelinated neocortical lesions in marmoset autoimmune encephalomyelitis mimic those in multiple sclerosis. Brain *128*, 2713-2721.

Sawcer, S., Hellenthal, G., Pirinen, M., Spencer, C. C., Patsopoulos, N. A., Moutsianas, L., Dilthey, A., Su, Z., Freeman, C., Hunt, S. E., *et al.* (2011). Genetic risk and a primary role for cell-mediated immune mechanisms in multiple sclerosis. Nature *476*, 214-219.

Storch, M. K., Bauer, J., Linington, C., Olsson, T., Weissert, R., and Lassmann, H. (2006). Cortical demyelination can be modeled in specific rat models of autoimmune encephalomyelitis and is major histocompatability complex (MHC) haplotype-related. J Neuropathol Exp Neurol *65*, 1137-1142.

Storch, M. K., Stefferl, A., Brehm, U., Weissert, R., Wallstrom, E., Kerschensteiner, M., Olsson, T., Linington, C., and Lassmann, H. (1998). Autoimmunity to myelin oligodendrocyte glycoprotein in rats mimics the spectrum of multiple sclerosis pathology. Brain Pathol *8*, 681-694.

Swanberg, M., Lidman, O., Padyukov, L., Eriksson, P., Akesson, E., Jagodic, M., Lobell, A., Khademi, M., Borjesson, O., Lindgren, C. M., *et al.* (2005). MHC2TA is associated with differential MHC molecule expression and susceptibility to rheumatoid arthritis, multiple sclerosis and myocardial infarction. Nat Genet *37*, 486-494.

Wall, M., Southwood, S., Sidney, J., Oseroff, C., del Guericio, M. F., Lamont, A. G., Colon, S. M., Arrhenius, T., Gaeta, F. C., and Sette, A. (1992). High affinity for class II molecules as a necessary but not sufficient characteristic of encephalitogenic determinants. Int Immunol *4*, 773-777.

Weiner, H. L. (2009). The challenge of multiple sclerosis: how do we cure a chronic heterogeneous disease? Ann Neurol *65*, 239-248.

Weissert, R., de Graaf, K. L., Storch, M. K., Barth, S., Linington, C., Lassmann, H., and Olsson, T. (2001). MHC class II-regulated central nervous system autoaggression and T cell responses in peripheral lymphoid tissues are dissociated in myelin oligodendrocyte glycoprotein-induced experimental autoimmune encephalomyelitis. J Immunol *166*, 7588-7599.

Weissert, R., Lobell, A., de Graaf, K. L., Eltayeb, S. Y., Andersson, R., Olsson, T., and Wigzell, H. (2000). Protective DNA vaccination against organ-specific autoimmunity is highly specific and discriminates between single amino acid substitutions in the peptide autoantigen [In Process Citation]. Proc Natl Acad Sci U S A *97*, 1689-1694.

Weissert, R., Svenningsson, A., Lobell, A., de Graaf, K. L., Andersson, R., and Olsson, T. (1998a). Molecular and genetic requirements for preferential recruitment of TCRBV8S2+ T cells in Lewis rat experimental autoimmune encephalomyelitis. J Immunol *160*, 681-690.

Weissert, R., Wallstrom, E., Storch, M. K., Stefferl, A., Lorentzen, J., Lassmann, H., Linington, C., and Olsson, T. (1998b). MHC haplotype-dependent regulation of MOG-induced EAE in rats. J Clin Invest *102*, 1265-1273.

Wucherpfennig, K. W., Sette, A., Southwood, S., Oseroff, C., Matsui, M., Strominger, J. L., and Hafler, D. A. (1994). Structural requirements for binding of an immunodominant myelin basic protein peptide to DR2 isotypes and for its recognition by human T cell clones. J Exp Med *179*, 279-290.

Zamvil, S. S., Mitchell, D. J., Moore, A. C., Kitamura, K., Steinman, L., and Rothbard, J. B. (1986). T-cell epitope of the autoantigen myelin basic protein that induces encephalomyelitis. Nature *324*, 258-260.

Zhao, W., Wegmann, K. W., Trotter, J. L., Ueno, K., and Hickey, W. F. (1994). Identification of an N-terminally acetylated encephalitogenic epitope in myelin proteolipid apoprotein for the Lewis rat. J Immunol *153*, 901-909.

2

Studies on the CNS Histopathology of EAE and Its Correlation with Clinical and Immunological Parameters

Stefanie Kuerten[1], Klaus Addicks[1] and Paul V. Lehmann[2,3]
[1]*University of Cologne*
[2]*Case Western Reserve University*
[3]*Cellular Technology Limited*
[1]*Germany*
[2,3]*USA*

1. Introduction

Multiple sclerosis (MS) is one of the most difficult to diagnose neurological diseases because its clinical manifestations are highly variable and the disease course also shows unpredictable individual patterns. We are far from understanding the complexities that underlie this variability, but certain patterns clearly emerge. First, it has become clear that different genetic backgrounds will lead to different manifestations of an autoimmune T cell attack on the central nervous system (CNS) (Hoppenbrouwers and Hintzen, 2010). It is also clear that differences in the CNS antigen-specificity of the T cell response can result in a differential involvement of anatomical regions of the CNS (Berger et al., 1997; Kuerten et al., 2007). Differences in lesion localization are a typical feature of MS, termed dissemination in space, and are likely to cause heterogeneity in clinical symptoms. There is evidence for the prevalence of either T cell-/macrophage- or antibody-/complement-mediated CNS demyelination versus a primary oligodendrogliopathy in MS patients (Lucchinetti et al., 2000). While the patterns of demyelination remain the same in individual patients over time, heterogeneity is evident when comparing different patients (Lucchinetti et al., 2000). In addition to these rather defined parameters of CNS histopathology (termed "pattern I-IV" by Lucchinetti et al., 2000) there are dynamic elements of the inflammatory cascade that can result in interindividual variations of disease progression. Among these are the extent of antigen determinant spreading (Lehmann et al., 1992; McRae et al., 1995), the prevalence of antigens in different CNS regions to which the spreading occurs (Targoni et al., 2001) as well as the rate at which regulatory or compensatory reactions of the immune system surface to counterregulate the damage of the target organ (Kasper et al., 2007).

Due to the impossibility of obtaining CNS tissue samples from individual patients repeatedly over time, studies as to the pathogenesis of the human disease need to rely on suitable animal models. To study pathologic features of MS three main animal models are used: disease induction by toxic agents, viral models, and finally different types of experimental autoimmune encephalomyelitis (EAE). Toxic agents like the copper chelator cuprizone cause demyelination in the relative absence of inflammation or axonal damage. Lesions induced in the cuprizone model typically resemble primary oligodendrocyte dystrophy in MS patients, while lacking the characteristic T cell infiltrate. The cuprizone model has no autoimmune component. Still, it is well-suited to investigate principle features of de- and remyelination in the CNS (Kipp et al., 2009). Intracerebral inoculation of Theiler's murine encephalomyelitis virus (TMEV) is used to investigate how viral infections can induce CNS autoimmunity. After an early, subtle disease period, susceptible mouse strains develop brain and spinal cord inflammation, demyelination and axonal damage. The clinical course resembles that of chronic, progressive MS (Tsunoda et al., 2010). However, EAE remains the most intensively studied animal model of MS.

2. Experimental Autoimmune Encephalitis (EAE)

EAE was introduced by Thomas Rivers and his colleagues in the early 1930s (van Epps, 2005). Since then, it has been subject to elaborate studies (reviewed in Goverman and Brabb, 1996; Steinman, 1999; Hemmer et al., 2002). Animals studied initially included guinea pigs and rats, in particular the Lewis rat, but later also involved marmoset monkeys and mice – the latter being the dominant model organisms used nowadays (Gold et al., 2006). EAE can either be induced by active immunization with CNS antigens in adjuvant (active EAE) or by the passive transfer of encephalitogenic T cells (adoptive/passive EAE). In addition, spontaneous EAE models relying on transgenic animals exist (Fig. 1).

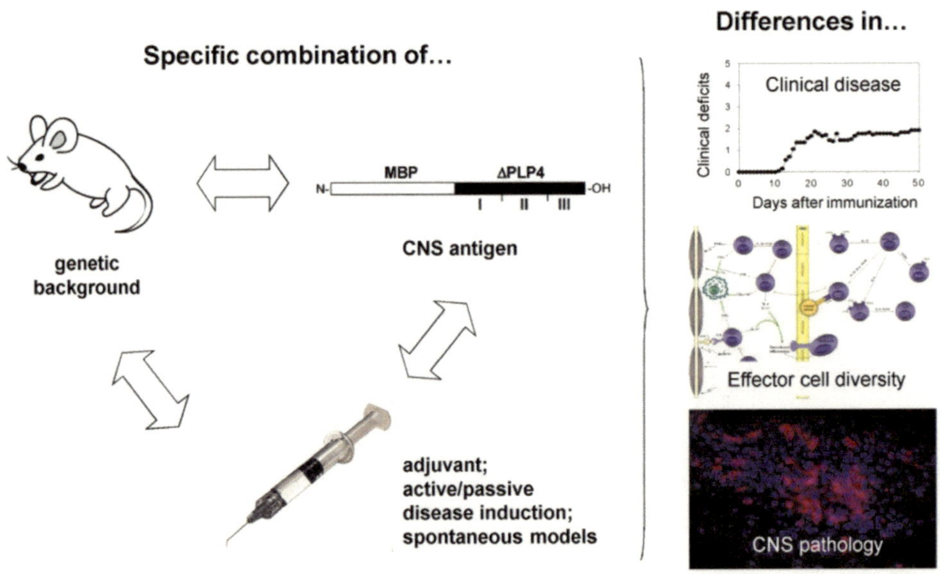

Fig. 1. The interplay between genetic background, disease triggering antigen and the mode of disease induction results in differences in EAE outcome.

Originally, whole spinal cord homogenate (SCH) (Einstein et al., 1962; Bernard & Carnegie, 1975; van Epps, 2005) was used for disease induction, before specific target antigens were defined. Early efforts to characterize the encephalitogenic antigen in SCH identified myelin basic protein (MBP) (Einstein et al. 1962; Martenson et al., 1970; Hashim et al., 1975) comprising approximately 30 - 40% of the proteins in the myelin sheath. H-2^u mice, in particular the B10.PL and PL/J strains are highly susceptible to MBP- or MBP peptide-induced EAE (Fritz et al., 1983 and 1985), while most common mouse strains, including C57BL/6 (B6) mice are resistant to MBP-induced disease (Bernard, 1976; Fritz and Zhao, 1996; Gasser et al., 1990; Skundric et al., 1994). The in-depth characterization of the B10.PL and PL/J model revealed highly restricted T cell responses to MBP involving a single immune dominant determinant and a limited usage of T cell receptor (TCR) chains (Zamvil et al., 1988; Urban et al., 1988; Kumar and Sercarz, 1994; Radu et al., 2000). Since similar findings were made in the Lewis rat (Burns et al., 1989), hopes emerged that such features could also apply to MS, providing therapeutic possibilities. These perspectives faded as diverse T cell repertoires were found in the proteolipid protein (PLP):139-151-induced EAE of SJL mice (Kuchroo et al., 1992) and after realizing that autoimmune T cell repertoires undergo determinant spreading (Lehmann et al., 1992; McRae et al., 1995; Jansson et al., 1995; Yu et al., 1996; Tuohy et al., 1999). Recent studies of antigen-specific autoantibodies in EAE have also provided for the diversification of the autoimmune response (Stefferl et al., 2000; Cross et al., 2001). PLP constitutes approximately 50% of the myelin proteins. As with MBP-induced EAE, only few strains were found to be susceptible to PLP-induced EAE. C57BL/6 mice were reported to be resistant (Tuohy, 1993; Mendel et al., 1995; Fritz and Zhao, 1996; Klein et al., 2000). The search for additional encephalitogenic antigens identified myelin oligodendrocyte glycoprotein (MOG) (Lebar et al., 1986; Mendel et al., 1995 and 1996; Schmidt, 1999), myelin associated glycoprotein (MAG) (Schmidt, 1999; Morris-Downes et al., 2002; Weerth et al., 1999), myelin oligodendrocyte basic protein (MOBP) (Schmidt, 1999; Holz et al., 2000; de Rosbo et al., 2004), oligodendrocyte-specific glycoprotein (OSP) (Morris-Downes, 2002), 2′,3′-cyclic nucleotide 3′ phosphodiesterase (CNPase) (Schmidt, 1999; Morris-Downes et al., 2002), β-synuclein (Mor et al., 2003) as well as S100β, which is not only expressed on astrocytes, but also in many other tissues including the eye, thymus, spleen and lymph nodes (Kojima et al., 1997; Schmidt, 1999).

Each combination of antigen with the respective susceptible strain and also considering the mode/protocol of disease induction results in a characteristic form of EAE (Goverman & Brabb, 1996; Steinman, 1999; Schmidt, 1999) (Fig. 1). The different EAE models show fundamental differences, however. For example, the MBP-induced disease in B10.PL and PL/J mice is monophasic: the mice completely recover after a single episode of short acute disease and become resistant to re-induction of EAE (Waxman et al., 1980). PLP peptide 139-151-induced EAE in SJL mice is remitting-relapsing (Hofstetter et al., 2002), while the disease elicited by MOG:35-55 in C57BL/6 mice is chronic (Eugster et al., 1999). In addition, the different EAE models involve differences in CNS histopathology and the role of antigen-specific antibodies, which will be described in detail below.

There have been extensive discussions regarding which antigen/strain combination provides the "best" EAE model for MS. The prevalent view is that none of them individually, but all of them jointly are best (Schmidt, 1999; van Epps, 2005; Hafler et al., 2005). MS does not seem to be a single disease entity, but rather involves a profound heterogeneity. As Vijay Kuchroo (Harvard University) once pointed out "each EAE model recapitulates a small piece of the human disease", thus facilitating the analysis of each single step disrupting immune competence leading finally to a severe autoimmune disease (van Epps, 2005; Steinman and Zamvil, 2006). EAE is an appropriate model for studies of basic mechanisms that underlie autoimmune pathology because, unlike in spontaneous autoimmune diseases, the autoantigen, the time point and the site of the ensuing autoimmune response is known, and the type of cytokine differentiation of the induced T cells can be directed (Forsthuber et al., 1996; Yip et al., 1999). Being able to control the above parameters as well as the ability to monitor the autoantigen-specific T cells in the course of the disease renders EAE suitable for studies aiming at defining the mechanisms of therapeutic interventions. Genetically-manipulated mice have been and will continue to gain increasing importance for such studies.

Traditionally, mechanistic studies have relied on the use of antibodies and on complex manipulations of mice. However, such treatments that can be applied to essentially any EAE model, do not necessarily permit unambiguous conclusions. For example, when a cell surface marker-specific antibody is injected to study the role of that molecule in EAE, the antibody might have a clinical effect on the disease, but it could be due to a multitude of mechanisms. The antibody could deplete the marker positive cells via the activation of complement, antibody-dependent cell-mediated cytotoxicity (ADCC) or apoptosis (Cebecauer et al., 2005), which in turn may be associated with a varying degree of inflammation causing unaccounted effects. Alternatively (or in addition) antibodies can inactivate or activate the marker positive cells, with variable bystander cell involvement. When antibodies are injected to study the role of a cytokine, the clinical effect seen can result from the neutralization of the cytokine, or on the contrary, from the prolongation of the half-life of that cytokine. For such reasons, the use of antibodies for mechanistic studies has frequently resulted in contradictory, inconclusive data (Dittmer and Hasenkrug, 1999; Silvera et al., 2001). The use of genetically-targeted mice, along with adoptive transfers of cells that express/do not express molecules of interest is increasingly becoming indispensable for mechanism-oriented studies, and the more this "tool box" expands the more powerful it will become. Most gene knock-out/knock-in mice have been generated on the 129 (*H-2^b*) background and backcrossed to *H-2* congenic C57BL/6 mice. Instead of having to move each new member of this ever expanding "toolbox" to the background of each EAE susceptible strain, it is much more effective to be able to study EAE in C57BL/6 mice. This is why MOG:35-55-induced EAE in C57BL/6 mice is increasingly becoming essential for mechanism-oriented studies - and why at the same time it is problematic to rely on this single EAE model for MS. To this end, we set out to establish and characterize additional EAE models for C57BL/6 mice. PLP protein-induced EAE has not been extensively studied; unlike the hydrophilic MBP molecule, PLP is highly hydrophobic and thus as a protein very difficult to utilize (Tuohy, 1993). PLP as an antigen for EAE induction established itself only after the encephalitogenic PLP peptide 139-151 had been identified for SJL mice (Kuchroo et al., 1992; Lehmann et al., 1992). Only recently, PLP peptide 178-191-elicited disease in C57BL/6 mice has been introduced (Tompkins et al., 2002), but this model still awaits thorough characterization. Encompassing most potential determinants of the two major myelin antigens, MBP and PLP, the MP4 fusion protein was generated as a drug candidate for MS (Elliott et al., 1996). MP4 contains the three hydrophilic loops of PLP (domains I, II and III; Fig. 2A), while the four hydrophobic transmembrane sequences have been excised. These hydrophilic domains constitute ΔPLP4 that has been linked to the 21.5 kD isoform of human MBP (Fig. 2B).

In SJL mice it has been shown that, when given under tolerogenic conditions, MP4 can prevent and revert EAE induced by MBP- and PLP-specific T cells (Elliot et al., 1996). It has also been shown that MP4 can induce EAE in SJL/J mice, and in another report (Jordan et al., 1999) MP4 was found to be encephalitogenic in marmoset monkeys. Our studies later demonstrated that MP4 was also capable of inducing EAE in C57BL/6 mice, thus introducing a much needed alternative to the MOG:35-55 and PLP:178-191 peptide model (Kuerten et al., 2006). Overall, there are few systematic studies as to whether different EAE models can reproduce distinct features of MS histopathology. One typical problem is that - as mentioned above - the induction of EAE requires the specific combination of genetic strain and CNS antigen. Yet, it is difficult to compare results obtained in different models since it is unclear, which outcome can be ascribed to the antigen and which one depends on the genetic background (Kuerten et al., 2009). It is therefore crucial to modify only one variable at a time, that is either the antigen or the genetic background. With the introduction of the MP4 model on the C57BL/6 background, the spectrum of models on this background covered all main antigens known from MS pathology: MOG, MBP and PLP. In addition, this background offers the possibility of performing genetic modifications, facilitating mechanistic studies.

In the following the characteristic histopathological features of MOG:35-55-, MP4- and PLP:178-191-induced EAE on the C57BL/6 background will be discussed and critically evaluated in the context of MS pathology considering the three hallmarks of MS pathology inflammation, demyelination and axonal damage.

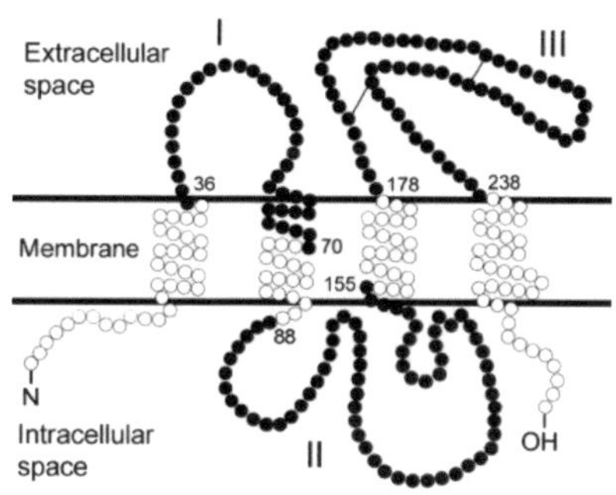

A

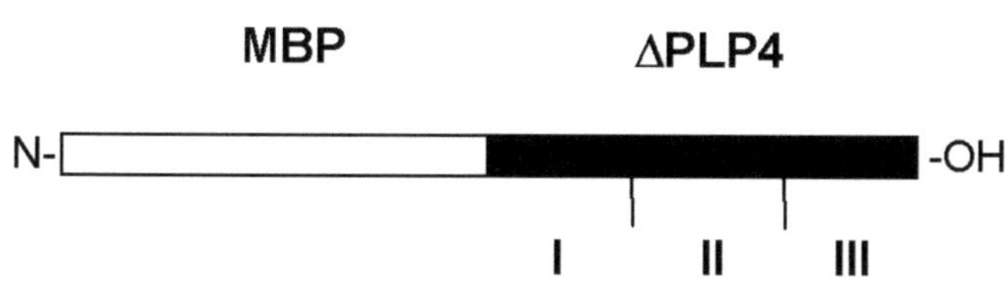

B

Fig. 2. The molecular structure of the MBP-PLP fusion protein MP4. (A) Structure of PLP. PLP is a transmembrane protein that consists of two extracellular (I and III) and an intracellular (II) hydrophilic domain, and four hydrophobic transmembrane sequences. (B) Structure of MP4. The three hydrophilic PLP domains have been fused to create ΔPLP4, which has been linked to the 21.5 kD isoform of human MBP.

3. Studies on the CNS histopathology of EAE and its correlation with clinical and immunological parameters

3.1 The CNS lesion topography and composition depends on the antigen used for immunization in models of C57BL/6 EAE

Inflammation is a feature of MS pathology that can essentially be reproduced in any EAE model. In principle, pathology is initiated when autoreactive T cells enter the CNS. Before, these cells need to be primed in the secondary lymphoid organs. In MOG:35-55- and PLP:178-191-induced EAE the antigen used for immunization is a single peptide and the autoreactive T cell response is directed against this peptide, while determinant spreading does not occur, which could include further determinants into the autoimmune response. The MP4 model, in contrast, is characterized by a multideterminant-specific $CD4^{+}$ T cell response and we have shown that there is no single dominant determinant being recognized in mice immunized with MP4 (Kuerten et al., 2006). Rather, the response seems to randomly target different determinants of the MP4 protein with interindividual variation in individual mice. The advantage of multideterminant specificity in the MP4 model resides in the fact that it may be used to better mirror the heterogeneity of the T cell response present in MS patients. It has been shown that there is not a single determinant targeted by the autoimmune response in MS. Differences do not only exist between individual patients, but also develop as disease progresses, since new determinants can be engaged into the immune response through determinant spreading (Tuohy et al., 1998). In patients this is a random process, which is highly unpredictable and can at least in part account for the kinetics of disease progression.

We assume that differences in the peripheral antigen-specific response have major implications on the subsequent CNS histopathology. In all models that we analyzed, infiltration of the cerebrum with focus on the meninges close to the hippocampal region occurred already in acute disease. In addition, in MP4- and PLP peptide-induced EAE inflammation of the spinal cord meninges was evident, while in the MOG peptide model inflammation extended into the parenchyma. Cerebellar infiltration was absent in the former two models, but pronounced in the latter in the acute stage of the disease. In chronic EAE (50 days after immunization), lesion distribution shifted towards the spinal cord and cerebellar parenchyma in the MP4 model, while it decreased remarkably in the cerebrum. In MOG peptide- and PLP peptide-induced EAE the lesion topography was comparable to the acute stage. Overall, CNS inflammation was time-dependent and dynamic only in the MP4 model, shifting from the cerebrum to the spinal cord and finally involving the cerebellum, thus allows the staging of the disease (Kuerten et al., 2007). MS is characterized by lesion dissemination in time and space, for which the MP4 EAE could serve as a valuable model. In contrast, the PLP and MOG model showed rather static inflammatory patterns that remained unchanged throughout the disease.

Next to differences in lesion topography we found differences in the cellular composition of CNS lesions. In particular, these differences pertained to the numbers of CNS infiltrating B cells.

3.2 The development of tertiary lymphoid organs (TLOs) in MP4-induced EAE of C57BL/6 mice

Studying the MP4, MOG peptide and PLP peptide model systematically early and late after immunization, we found that B cell infiltration was a common feature of the MP4 model, while in MOG peptide- and PLP peptide-induced disease B cells were scarce within the infiltrates (Kuerten et al., 2008). Moreover, it was striking that in MP4-induced disease B cells showed clustering, while in the other models they were scattered throughout the lesions (Fig. 3).

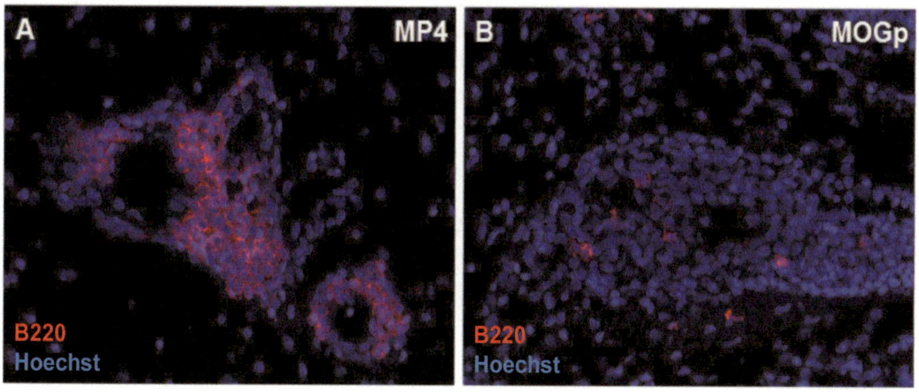

Fig. 3. Presence of CNS B cell aggregates in the MP4 model. C57BL/6 mice were immunized with 150 µg MP4 or 100 µg MOG peptide 35-55 (MOGp) in complete Freund's adjuvant (CFA). Pertussis toxin was given at 200 ng per mouse on the day of immunization and 48 h later. 35 days after disease onset mice were sacrificed, the CNS was removed and snap-frozen in liquid nitrogen. Seven µm thick cryostat sections were obtained from cerebrum, cerebellum and spinal cord and stained for the presence of B cells using B220 antibody. Representative cerebellar parenchymal infiltrate from a (A) MP4-immunized and a (B) MOG peptide 35-55-immunized mouse. Images are at 400x magnification and representative for 58 mice tested in at least eight independent experiments (Kuerten et al., 2008).

The presence of B cell aggregates could be indicative of the formation of ectopic foci of lymphoid tissue - termed tertiary lymphoid organs (TLOs) - in the MP4 model. Lymphoid organization of inflamed tissue can occur in the setting of chronic inflammation and is mainly directed by the expression of lymphotoxin ($LT\alpha_1\beta_2$) by activated B and T cells that interacts with the lymphotoxin-β receptor on stromal organizer cells (Aloisi & Pujol-Borrell, 2006). The structure of TLOs has been reported to be variable (Drayton et al., 2006) and there has been controversy as to the exact definition of a TLO. Overall, TLOs are thought to resemble lymph nodes (Aloisi & Pujol-Borrell, 2006; Deteix et al., 2010). Major features of TLOs include T cell/B cell compartmentalization, the presence of high endothelial venules (HEVs) that allow naive B and T cells to enter the tissue as well as the expression of lymphoid chemokines such as CXCL13, CCL12, CCL19 and CCL21. In addition, follicular dendritic cell (FDC) networks have been associated with TLO formation and occasionally germinal centers have been found. The presence of germinal centers in TLOs points to the fact that these structures are not solely epiphenomena that emerge in chronic inflammation, but that these organs can be functional and may influence disease progression in the setting of autoimmunity, for example by the production of high-affinity autoantibodies and by the facilitation of determinant spreading (Stott et al., 1998; Armengol et al., 2001; Sims et al., 2001). Somatic hypermutation, affinity maturation, immunoglobulin class switching and B cell receptor revision are all processes that take place in secondary lymphoid organs (SLOs) and there is increasing evidence that they can also occur in TLOs, probably contributing to the exacerbation of the chronic inflammatory state and also to the detachment from the immune response generated in SLOs. The formation of TLOs has been found under a variety of pathogenic circumstances including autoimmune diseases such as Sjögren's syndrome (Aziz et al., 1997; Salomonsson et al, 2003; Barone et al., 2005), autoimmune thyreoiditis (Armengol et al., 2001) and arthritis (Takemura et al., 2001; Shi et al., 2001), infectious diseases (Murakami et al., 1999; Mazzucchelli et al., 1999), tumors (Coronella et al., 2002; Nzula et al., 2003) and transplantation (Baddoura et al., 2005). In MS B cell aggregates have been identified in the meninges of patients with secondary-progressive MS (SP-MS). Owing to the expression of CXCL13, the presence of FDCs, proliferation (indicative of germinal center formation) and plasma cells these aggregates have been described to be ectopic B cell follicles (Serafini et al., 2004; Magliozzi et al., 2007). The presence of ectopic B cell follicles has been linked to a younger age at disease onset, irreversible disability and death in addition to more pronounced demyelination, microglia activation and loss of neurites in the cerebral cortex (Magliozzi et al., 2007). In a follow-up study, Serafini et al. provided evidence for an association between B cell follicle formation in the CNS and latent Epstein-Barr virus (EBV) infection, which they suggested to contribute to B cell dysregulation (Serafini et al., 2010). These data have important implications for the disease pathogenesis since they propose a histopathological correlate for sustained disease and its chronification as well as they strongly support the viral hypothesis of MS. However, it should be noted that other researchers failed not only to detect meningeal B cell follicles and an association between meningeal inflammation and cortical demyelination, but also the presence of EBV-infected cells in the CNS of MS patients (Kooi et al., 2009; Willis et al., 2009; Perferoen et al., 2010). The debate about the actual presence and relevance of B cell follicles/ectopic lymphoid tissue in MS has not yet been resolved.

Since research involving tissue from MS patients is restricted and problems emerge when it comes to defining the exact onset and further course of the disease, studies in EAE could help clarify the controversy about B cell follicles in the disease process. In EAE, disease onset and progression are clearly defined, which facilitates the correlation of ectopic lymphoid tissue development and clinical outcome. Yet, most traditional EAE models are independent of B cells. Among these models are (as mentioned above) the traditional MOG peptide 35-55- and PLP peptide 178-191-induced EAE in C57BL/6 mice and the PLP peptide 139-151-elicited disease in the SJL/J strain. Only the human MOG (Oliver et al., 2003) and a transgenic SJL model (Pöllinger et al., 2009) have been shown to comprise a pathogenic B cell response, but have so far not proven to be helpful when it came to studying lymphoid tissue formation in the CNS. Ongoing studies in our laboratory are currently dealing with a thorough analysis of the B cell aggregates found in the CNS of MP4-immunized mice. We have obtained first evidence that these aggregates indeed take over characteristics of TLOs including B cell and T cell compartmentalization (Fig. 4).

In addition to B cell/T cell compartmentalization, CNS B cell aggregates in MP4-induced EAE showed further characteristic features of TLO formation including the presence of HEVs, FDCs and chemokine expression (Fig. 5).

Having established that MP4 immunization induces TLO formation in C57BL/6 mice, specific questions addressed in future studies will be concerned with the topography and kinetics of TLO formation in MP4-induced EAE, the correlation with the clinical disease outcome and cortical pathology. In addition, we aim at elucidating whether TLOs in MP4-

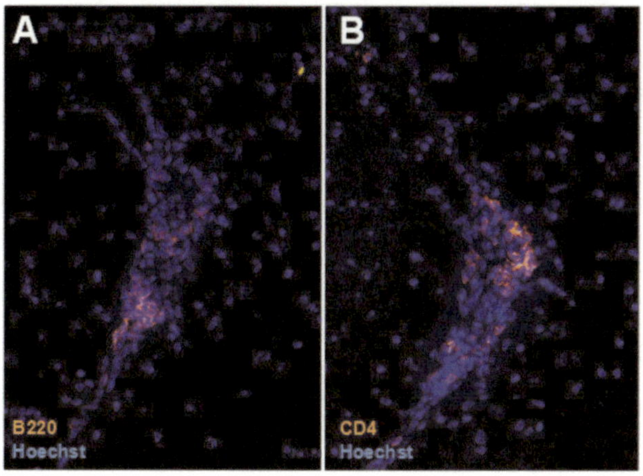

Fig. 4. Presence of B cell/T cell compartmentalization within a CNS B cell aggregate in MP4-induced EAE. C57BL/6 mice were immunized with 150 µg MP4 in CFA. Pertussis toxin was given at 200 ng per mouse on the day of immunization and 48 h later. 52 days after disease onset mice were sacrificed, the CNS was removed and snap-frozen in liquid nitrogen. Seven µm thick cryostat sections were obtained from the cerebrum, cerebellum and spinal cord and stained for the presence of B cells (A) and CD4^{+} T cells (B). A representative infiltrate in the cerebral meninges is shown. Images are at 400x magnification and representative for a total of 24 mice tested in six independent experiments (Kuerten et al., 2011c).

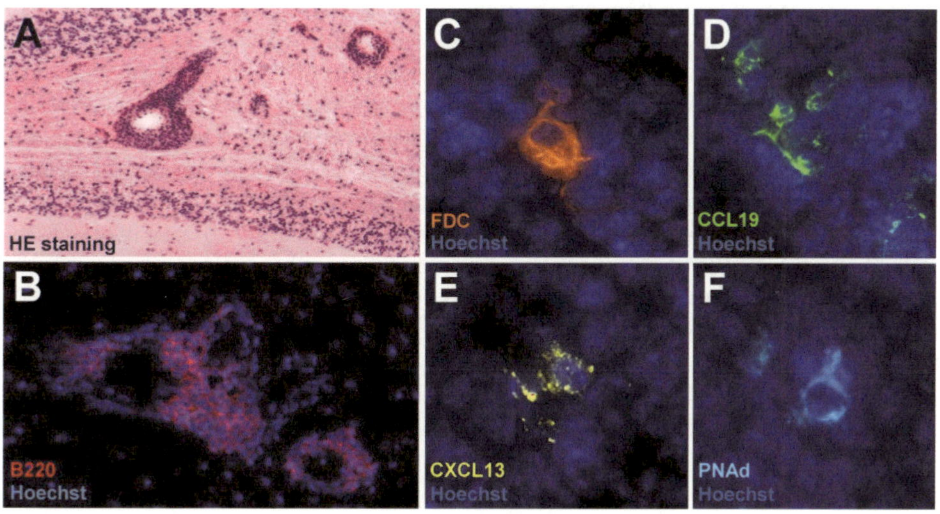

Fig. 5. Characteristics of CNS TLO formation in MP4-induced EAE. C57BL/6 mice were immunized with 150 µg MP4 in CFA. Pertussis toxin was given at 200 ng per mouse on the day of immunization and 48 h later. 35 days after disease onset mice were sacrificed, the CNS tissue was removed and snap-frozen in liquid nitrogen. Seven µm thick cryostat sections were obtained from cerebrum, cerebellum and spinal cord and stained for the presence of B cells (A,B), FDCs (C), CCL19 (D), CXCL13 (E) and PNAd expressed in HEVs (F). A representative cerebellar parenchymal infiltrate from a MP4-immunized is shown. (A) is at 100x, (B) at 200x and (C-F) are at 630x magnification. The images are representative for a total of 24 mice tested in six independent experiments (Kuerten et al., 2011c).

induced EAE are functional - here our focus will be laid onto the role of TLOs for determinant spreading of the T cell and B cell response. Determinant spreading can substantially contribute to the chronification and diversification of the immune response in autoimmune disease (Lehmann et al., 1992; McRae et al., 1995; Tuohy et al., 1998). Revealing TLOs as structures responsible for determinant spreading will underline the importance of studies that evaluate TLOs as therapeutic targets. The therapeutic disruption of TLO formation could be a potential means to slow down or ideally prevent disease progression in multiple sclerosis and other autoimmune disorders.

3.3 The immunopathology of MP4-induced EAE is autoantibody-dependent

We have shown that the involvement of B cells in the MP4 model is not restricted to the infiltration of these cells into the CNS and the formation of TLOs, but that it additionally includes the production of antibodies by autoreactive B cells. Antibodies have been noted in a variety of EAE models, directed against MOG, MBP and PLP (Sadler et al., 1991; Lyons et al., 2002). However, the presence of antibodies in the serum of immunized mice does not directly imply their pathogenicity. Among others, antibodies directed against MOG peptide 35-55 or rat MOG protein have been shown to be non-pathogenic (Oliver et al., 2003; Marta et al., 2005), while antibodies against human MOG protein have been associated with demyelination (Lyons et al., 2002; Oliver et al., 2003). Immunization with MP4 clearly triggered the production of MP4-specific antibodies (Kuerten et al., 2011a). Antibodies reactive to MP4 were evident as early as 15 days after immunization. The MP4-specific antibody response reached a plateau around day 50 after immunization. The MP4-specific antibodies were of the IgG1 and IgG2a isotype, with IgG1 apparently prevailing, but the difference did not reach statistical significance. In addition, these MP4-specific antibodies proved to be myelin-reactive. C57BL/6 mice were immunized with PBS in CFA or MP4 in CFA, and B cell-deficient μMT mice were immunized with MP4. On day 40 after immunization mice were bled and serum was isolated. Consecutively, frozen longitudinal spinal cord sections obtained from naïve untreated C57BL/6 wild-type mice were incubated with the serum to evaluate myelin reactivity. Staining of the myelin sheath was only evident when incubating spinal cord sections with serum obtained from MP4-immunized wild-type C57BL/6 mice, while staining was absent when using control serum from PBS/CFA-immunized or MP4-immunized μMT mice. In the following, we demonstrated that these myelin-reactive antibodies, however, were not able to mediate pathology on their own. When immunized with MP4, the two congenic B cell-deficient mouse strains μMT and J_HT did not develop EAE, emphasizing the role of B cells in the MP4 model (Kuerten et al., 2011a). Transfer of MP4-reactive serum into B cell-deficient mice did not revert this resistance. The permeabilization of the blood-brain barrier (BBB) by injection of pertussis toxin in parallel to the serum transfer did not result in clinically and/or histologically evident EAE either. Merely the additional immunization of B cell-deficient mice with MP4 restored disease to the level of the wild-type mice. In this set of experiments, the serum transfer protocol as established by Lyons et al., 2002 was used. MP4-reactive serum was isolated from C57BL/6 donor mice on days 30, 50, 70 and 90 after immunization. MP4 reactivity was tested by ELISA and myelin binding capacity by incubating spinal cord tissue from naïve untreated C57BL/6 mice with each particular serum batch. Only serum that tested positive in both ELISA and immunohistochemistry was used for subsequent transfer. Serum was transferred four times, that is on days 0, 4, 8 and 12 adding up to a total of 600 μl of serum transferred into each recipient mouse (150 µl of serum were transferred on each time point, diluted to a total of 500 µl injection volume in sterile PBS).

The mechanism by which MBP/PLP-specific antibodies assert the EAE sensitizing effect in MP4-immunized mice is unclear. We have shown that MP4-induced antibodies stain myelin on tissue sections. Because the sections go through the myelin sheath, such staining, however, does not permit the distinction between intra- or extracellular binding of the antibodies. When studying CNS sections of mice undergoing EAE antibody depositions were seen in lesions only, while being absent in parts of the CNS, in which cellular infiltration and tissue damage was not evident. This finding supports the hypothesis that autoantibodies can assert local effects only in synergy with T cell-induced inflammation, which disrupts the blood-brain barrier (BBB) and also permits complement components to enter. Fig. 6 shows colocalization of demyelination, antibody and complement depositions in a spinal cord lesion induced by immunization with MP4. The data point to the involvement of the complement system in the MP4 model.

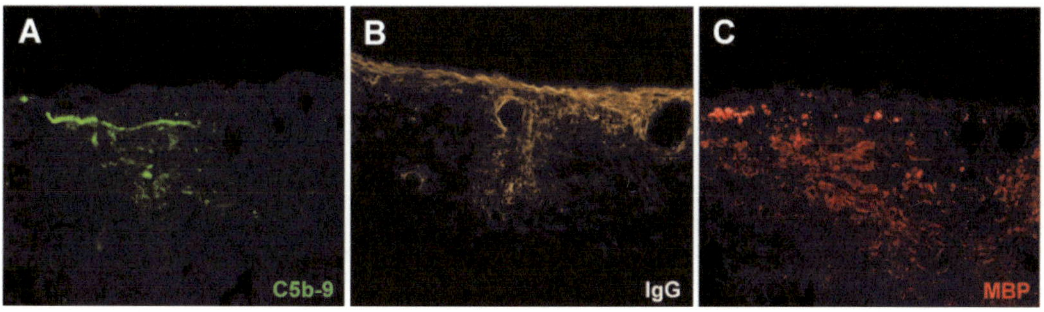

Fig. 6. Colocalization of demyelination and antibody/complement depositions in CNS lesions of MP4-immunized mice. C57BL/6 mice were immunized with 150 µg MP4 in CFA. Pertussis toxin was given at 200 ng per mouse on the day of immunization and 48 hours later. Mice were sacrificed on day 35 after immunization, the spinal cord was removed and 7 µm cryostat sections were obtained. The tissue was stained for the deposition of C5b-9 (the membrane attack complex MAC) (A) and IgG (B) in colocalization with demyelination (C). Results are representative for a total of n = 6 mice analyzed in two independent experiments. Images are at 100x magnification and refer to a mouse with a clinical score of 2.5 (Kuerten et al., 2011d).

While PLP may be the primary target of the autoimmune response due to its extracellular domains, MBP might be involved in later disease stages once myelin breakdown has occurred. It will be of interest to evaluate the relevance of various myelin antigens in different disease stages. To this end, future studies will require working with individual domains of MP4. In addition, the generation of monoclonal antibodies against individual domains of PLP and the MBP molecule will be needed to further dissect the fine-specificity of MP4-specific antibody action.

3.4 Demyelination, axonal damage and gray matter pathology in C57BL/6 EAE

Inflammation is known to be the characteristic attribute of the acute stages of the disease. However, differences have been observed when comparing different EAE models. In MOG peptide 35-55 and PLP peptide 178-191 EAE of C57BL/6 mice three months after immunization only few disseminated infiltrating cells, in particular CD4+ T cells and macrophages, were present in the tissue. In the MP4 model, inflammation was more sustained: while inflammatory foci decreased over time in spinal cord and cerebrum, cerebellar infiltration was a prevalent feature of chronic EAE three months after immunization. Besides inflammation, demyelination and axonal damage are considered to be hallmark features of MS and therefore also need to be addressed in studies of the animal model. In our follow-up study we investigated the degree of demyelination, axonal damage and motor neuron pathology in MP4-, MOG peptide and PLP peptide-induced EAE of C57BL/6 mice. We demonstrated that major differences between the three models resided in (i) the region-/tract-specificity and disseminated nature of spinal cord degeneration, (ii) the involvement of motor neurons in the disease and (iii) the extent and kinetics of demyelination.

Not many studies have dealt with a systematic investigation of the differential involvement of spinal cord fiber tracts in EAE. The murine spinal cord can be subdivided into three main fiber tract systems - the anterolateral tract (that carries on pain, temperature and crude touch sensation), the dorsal tract (that transmits fine touch sensation) and the pyramidal tract (responsible for motor function). Most reports either do not provide information about which tract has actually been analyzed, or reports focus on the anterolateral tract. However, the clinical symptoms evident in the mice cannot solely be explained by anterolateral tract pathology. Mice typically present with a floppy tail initially that advances into an ascending paralysis as the disease progresses. To this end, any correlation between clinical deficits and CNS histopathology also needs to involve studies of pyramidal tract and motor neuron alterations.

Our data demonstrate that the anterolateral tract was affected in all mice in MOG peptide 35-55-, PLP peptide 178-191- and MP4-induced EAE and to a similar extent. The dorsal tract showed more gradual pathology in acute EAE, but was targeted in 100% of mice in chronic EAE. While a clinical assessment of sensory deficits in mice is hard to perform, our data show that EAE is suitable to study the pathology of this neurological quality (Kuerten et al., 2011b). Autoimmune encephalomyelitis has originally been believed to be a "white matter disease". However, reports exist that suggest additional pathologic changes in the gray matter including loss and/or atrophy of motor neurons (Bannerman et al., 2005; Fisher et al., 2008; Derfuss et al., 2009; Rudick & Trapp, 2009). In MS patients CNS gray matter has been shown to be affected at multiple sites covering the basal ganglia, the hippocampus (Derfuss et al., 2009), spinal cord and the cortex (Kidd et al., 1999; Bo et al., 2003; Wegner et al., 2003; Kutzelnigg et al., 2005; Kutzelnigg et al., 2007). In addition, gray matter pathology has been shown to reflect functional disability better than the extent of white matter plaque formation (Wegner et al., 2003; Fisher et al., 2008). Consecutively, contactin-2 has been suggested as gray matter target antigen for the autoimmune T cell response (Derfuss et al, 2009). Motor neuron pathology as one aspect of gray matter disease has not been investigated extensively yet. Vogt and colleagues (Vogt et al., 2009) were able to demonstrate that in MS patients compound muscle action potential amplitudes and motor unit numbers were decreased compared to controls subjects, which was indicative of lower motor neuron loss. The data were confirmed by high-precision unbiased stereological quantification of spinal cord neurons in *post-mortem* MS tissue. In this material, T cells secreting TNF-related apoptosis inducing ligand TRAIL were found in close vicinity to apoptotic neurons (Vogt et al., 2009).

Studies as to whether gray matter pathology is a characteristic feature of different EAE models are scarce. Bannerman et al. (2005) have reported motor neuron atrophy in MOG peptide 35-55-induced EAE of C57BL/6 mice by staining of hypophosphorylated neurofilament H with SMI-32 antibody. In our study we extended the analysis of neurofilament H phosphorylation patterns to also include PLP peptide- and MP4-induced EAE. The data show that in PLP peptide-induced EAE the motor neuron phenotype did not show any signs of atrophy, while in the MP4 and MOG peptide model significant alterations were evident.

We agree with findings presented by Bannerman et al. (2005) in the context of MOG:35-55-induced EAE suggesting motor neuron atrophy evidenced by significantly diminished SMI-32 reactivity. In addition, this group reported that such abnormalities were much less prominent by 14 weeks post immunization. We followed up on spinal cord pathology for about 12 weeks. We share the regression of motor neuron phenotype alterations in MOG peptide-induced EAE and we delineate that in the MP4 model motor neuron pathology seems to be more persistent (similar to the chronic demyelination and axonal damage in this model) (Kuerten et al., 2011b).

Loss of spinal cord motor neurons has been reported in MBP-induced EAE of Lewis rats (Smith et al., 2000). In MOG peptide 35-55-induced EAE of C57BL/6 motor neuron loss has not been noted (Bannerman et al., 2005). On the one hand, no TdT-mediated dUTP-biotin nick end labeling (TUNEL) positive neurons were observed in any of the mice analyzed on days 14, 21 or 98 post immunization (Bannerman et al., 2005). On the other hand, there was also no difference in the densities of motor neurons counted in cross sections of L5,L6 spinal cords in MOG peptide-EAE and CFA control mice on day 98 after immunization. Finally, the analysis of L5,6-innervated skeletal muscles did not show any muscle atrophy, angulated skeletal muscle fibers no fiber type-specific grouping in MOG peptide-immunized animals on day 98 after immunization, indicating that these muscle fibers were neither denervated nor reinnervated by axonal collateral sprouting (Bannerman et al., 2005). We also performed analysis of apoptosis in motor neurons in MOG peptide 35-55-, PLP peptide 178-191- and MP4-induced EAE staining for TUNEL or caspase 3. In accordance with Bannerman et al., 2005 we did not observe motor neuron apoptosis in the three models (unpublished data). The cause of the alteration of the motor neuron perikaryal phosphorylation in the MOG peptide and MP4 model remains to be elucidated. Previous studies have shown that increased phosphorylation of neuronal neurofilament H can be induced by an increase in the extracellular concentration of glutamate (Ackerley et al., 2000). Glutamate excitoxicity is believed to play an important pathogenic role in EAE and MS (Hardin-Pouzet et al., 1997; Matute et al., 2001; Werner et al., 2001) and could also be of importance for causing neuronal damage. Future studies clearly need to address the mechanisms underlying motor neuron alterations in EAE and possibly also MS. To further assess alterations in the motor neuron phenotype we are currently conducting ultrastructural analysis using electron microscopy. Our preliminary data show that in addition to changes in the neurofilament H phosphorylation patterns, motor neuron pathology can include nuclear membrane dissolution, vacuolization of the cytoplasm and a decrease in synaptic densities (Fig. 7). Only mild motor neuron degeneration was evident in MOG:35-55-immunized mice, characterized by an increased number of intracytoplasmic vacuoles and slight nuclear changes that encompassed an irregular undulation of the nuclear membrane. MP4-induced EAE in contrast led to more severe nuclear membrane

A	overall degeneration	vacuoles	rough ER	nuclear changes	synapses per mm
control			++		+++
MOG mild	+	+	++	+	++
MOG severe	+	+	++	+	++
MP4 mild	++	++	++	+++	++
MP4 severe	++	++	+	+++	++

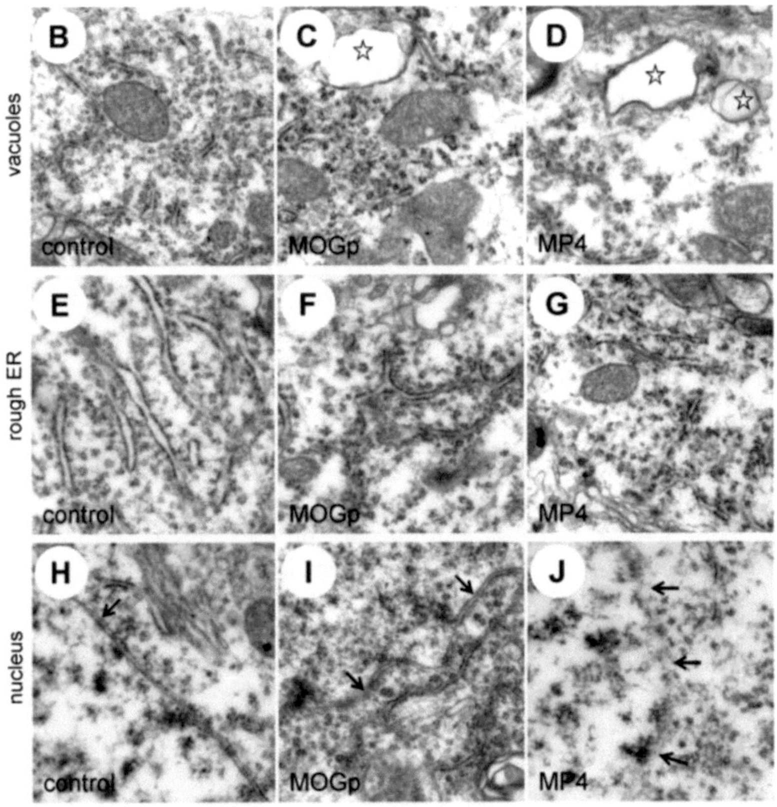

Fig. 7. MP4-induced EAE displays severe motor neuron pathology. C57BL/6 mice were immunized with 150 µg MP4 or 100 µg MOG:35-55 in CFA. Pertussis toxin was given at 200 ng per mouse on the day of immunization and 48 hours later. (A) Motor neuron pathology was assessed using a semi-quantitative scoring system, which considered the overall degree of motor neuron degeneration, the occurrence of intracytoplasmic vacuoles (B-D), rough ER (E-G) and nuclear changes (H-J) as well as the number of synapses per mm. Stars in panels (F) and (G) designate vacuoles, the arrows in panels (I) and (J) indicate the nuclear membrane. All images are at 12.000x magnification. The data refer to n = 6-8 mice tested in each group and tested in at least three independent experiments. Data obtained from EAE mice were compared to data from untreated control mice. Mild EAE encompassed the clinical scores 0.5-2, severe EAE referred to clinical scores > 2. The mean clinical score in the MOG peptide and MP4 group was similar both in mild and severe EAE, respectively (1.40 ± 0.22 versus 1.35 ± 0.03 in mild MOG peptide versus MP4-induced EAE with p = 0.593 and 2.58 ± 0.20 versus 2.66 ± 0.28 in severe MOG peptide versus MP4-induced EAE with p = 0.582).

defects up to complete nucleic resolution. The extent of rough endoplasmic reticulum (rER) alterations and the number of synapses remained largely unchanged compared to the control group. Only in severe MP4-induced EAE we noted beginning resolution of the rER. The number of synapses per mm was decreased in both MOG:35-55- and MP4-induced EAE compared to controls (Gruppe et al., 2011).

Besides motor neuron dysfunction, damage to the pyramidal tract can be responsible for the development of motor deficits. What was intriguing to see was the fact that in our light microscopic analysis of methylene blue-stained transverse spinal cord sections in the MP4 model almost exclusively motor neuron perikaryal disturbances were evident, while in the MOG and PLP peptide model the pyramidal tract equally showed degeneration (Kuerten et al., 2011b). The perikaryal disturbances were characterized by an increase in staining intensity of the Nissl substance. These microscopically visible changes could be due to a transient increase in Nissl substance due to a loss of trophic input as a consequence of EAE-induced motor neuron dendritic pathology (Zhu et al., 2003; Bannerman et al., 2005). Considering our ultrastructural data that showed a decrease of rough ER (and thus Nissl substance) over time, it is - however - more likely to favour the alternative hypothesis that the light microscopic picture is due to rough ER dissolution/fragmentation.

However, at this point it should be noted that a conclusive statement as to the pathologic changes in individual fiber tracts cannot be made without further ultrastructural analysis. Similar to our analysis of motor neuron pathology we are currently also conducting electron microscopic studies of pyramidal tract pathology. Our data indicate that pyramidal tract pathology also occurs in the MP4 model including demyelination and axonopathy, however the extent of pyramidal tract pathology seems to be less severe compared to the MOG peptide 35-55 model, which could explain the differences observed in our light microscopic analysis.

Next to the differential targeting of spinal cord fiber tracts, another differential histopathological feature applied to the extent and kinetics of demyelination in MP4-, MOG peptide 35-55- and PLP peptide 178-191-induced EAE. Chronic demyelination in the MP4 model was opposed to only transient or absent myelin pathology in MOG peptide and PLP peptide EAE. Considering these data in the context of what we have discussed above, the MP4 and MOG peptide model could help reproduce distinct demyelinative patterns. MP4-induced EAE may be a valuable tool for studying myelin pathology that relies on B cells/autoantibodies and complement activation versus demyelination primarily caused by T cells and macrophages in the MOG peptide model. In addition, the comparison of these models to the non-demyelinating disease induced by PLP peptide 178-191 may give insight into factors that actually initiate and maintain the myelin attack dependent on the antigen that is the prime target. When evaluating the activity of the lesion by staining for major histocompatibility (MHC) class II we noted the transition from active plaque regions with high expression of MHC II to chronic still demyelinated, but "MHC II low" lesions in MP4-induced EAE. The presence of highly inflammatory lesions later turning into chronic burnt-out demyelination is a hallmark of MS (Trapp et al., 1996), which still needs further investigation, for which the MP4 model may be a suitable tool.

Finally, besides all these differences in CNS histopathology, the analysis of the clinical outcome of EAE induced by MP4, MOG peptide 35-55 and PLP peptide 178-191 showed major similarities in the course and severity of the disease (Kuerten et al., 2007). While these similarities were clearly opposed to the differential patterns of demyelination and regional spinal cord pathology that we have just discussed, we found the extent of axonal damage to be highly comparable in the three models (Fig. 8). Therefore, in accordance with data obtained in MS patients (Trapp et al., 1998), we propose axonal injury as the main structural-morphological correlate causing irreversible functional deficits, but we acknowledge that future studies on the ultrastructural level are still needed to support this notion.

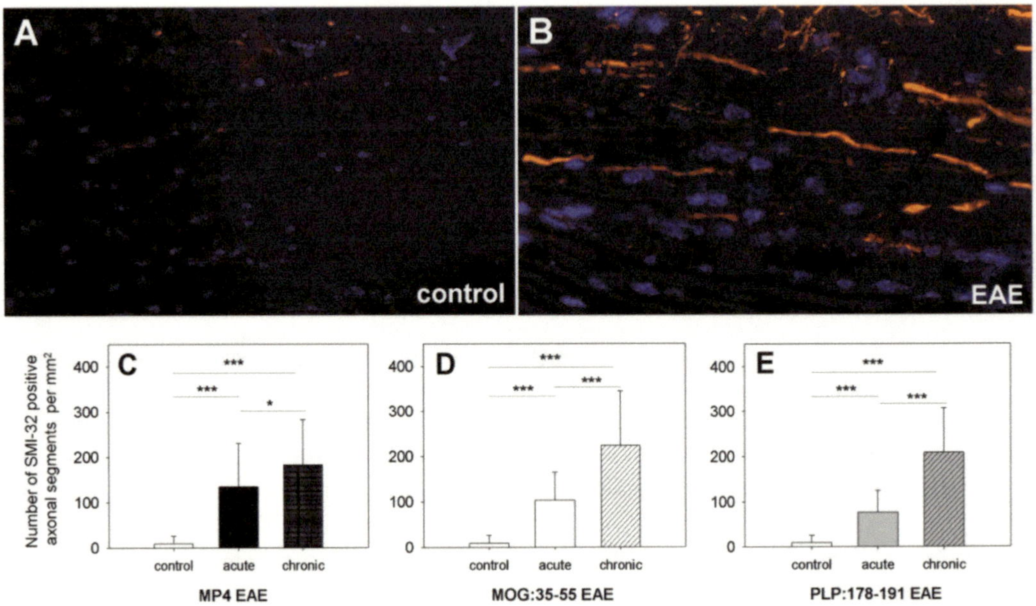

Fig. 8. Increasing axonal damage in the course of MP4-, MOG:35-55-, and PLP:178-191-induced EAE. C57BL/6 mice were immunized with 150 µg MP4, 100 µg MOG peptide 35-55 or 200 µg PLP peptide 178-191. Pertussis toxin was given at 200 ng per mouse on the day of immunization and 48 h later. Longitudinal 7 µm thick frozen spinal cord sections were stained with SMI-32 antibody. Representative images are shown for spinal cord sections from PBS/CFA control-immunized C57BL/6 mice (control) (A) and EAE tissue, here taken from a MP4-immunized mouse three months after onset of the disease with a score of 2.5 (B). Panels (C-E) display the mean number of SMI-32 positive axonal segments per mm^2 + SD in each model on the peak of acute EAE (approximately day 15 in the MP4 model, day 20 in the MOG peptide 35-55 model and day 30 in PLP peptide EAE) and in the chronic stage of EAE three months after immunization compared to control-immunized mice. The data refer to n = 10 mice tested in each group in three independent experiments. All images are at 400x magnification. * $p = 0.05$, *** $p < 0.001$.

In conclusion, while all three models - MP4-, MOG 35-55- and PLP:178-191-induced EAE in C57BL/6 mice - clinically display chronic disease with comparable severity and course, the CNS histopathology underlying the functional clinical deficits can follow differential patterns. Our data suggest the use of MP4-, MOG peptide 35-55- and PLP peptide 178-191-induced EAE on the C57BL/6 background as a reasonable strategy for reproducing distinct features of CNS pathology fuelling work towards a better understanding of MS diversity.

4. Future implications

Eventually, our experimental approach is meant to go beyond the investigation whether different EAE models can be reflective of different patterns of CNS pathology. A clinically highly relevant question deals with the correlation between CNS histopathology and the clinically evident outcome in each individual patient. While axonal damage is assumed to cause irreversible deficits, due to the limited access to human CNS samples and the impossibility of obtaining samples repeatedly over time there are few reports so far directly analyzing in which way or how the degree of CNS pathology defines the clinical severity. The kinetics of lesion development in patients have mostly been assessed by MRI (Inglese et al., 2011; Sicotte et al., 2011). While this approach allows to establish a correlation between overall lesion load and clinical disease severity, it does not provide information about the mechanisms underlying lesion development. Even further, it would be clinically highly relevant to investigate whether there is a correlation between CNS pathology and peripherally measurable immunological parameters.

In an initial study we set out to analyze the correlation between the magnitude of the interferon-γ(IFN-γ)/interleukin-17 (IL-17) antigen-specific response in the blood and the clinical course of the disease in two independent EAE models: the remitting-relapsing PLP peptide-induced EAE of SJL/J mice and the chronic disease of C57BL/6 mice immunized with MOG peptide 35-55. To this end, we established an experimental technique that worked with low amounts of blood, thereby permitting longitudinal and repeated testing of mice. The technique we used relied on a double-color enzyme-linked immunospot technique (ELISPOT)-based test system, for which as little as 150 µl of murine blood sufficed. We then tested mice repeatedly over the time course of EAE to establish the kinetics of the antigen-specific blood IFN-γ and IL-17 T cell response in both the MOG peptide 35-55/C57BL/6 and the PLP peptide 139-151/SJL model. In sum, the data delineate that the dynamic course of EAE in the SJL/J model was closely reflected by dynamics in the blood T cell compartment, while chronic EAE in the C57BL/6 model was mirrored by a relatively stable antigen-specific T cell response (Kuerten et al., 2010).

As we learn more about the antigens that are actually targeted in MS, our approach may serve as a valuable approach towards more efficient prognostic and diagnostic options in patients. Despite remarkable scientific effort, MS has remained highly unpredictable and a suitable biomarker for the disease has not been found (hallmark studies are summarized in Galboiz & Miller 2002; Rinaldi & Gallo, 2005 and Reindl et al., 2006). To date, there is no possibility of determining whether a patient presenting with clinically-isolated syndrome (CIS) or radiologically-isolated syndrome (RIS) will develop definite MS. There is also no possibility of predicting the course of disease in MS patients, and in particular whether and when a patient in clinical remission will develop a relapse. In addition, it is assumed that several subpopulations of MS exist, and the contribution of CNS antigen- and in particular myelin-reactive B and T cells differs in these subpopulations. While in a majority of patients autoreactive T cells are detectable in CNS demyelinative lesions, there are also subpopulations of patients, in which a primary oligodendrogliopathy is evident. This difference in response is likely to result from differences in immune pathogenesis underlying the disease. So far, there are also no methods available that permit the prediction of treatment responsiveness in MS patients.

In an ongoing study in collaboration with the Department of Neurology, University Hospitals of Cologne, we perform measurements of CNS antigen/myelin-specific T and B cells in the blood after *in vitro* stimulation of these T cells with CNS/myelin antigen. It is our aim to determine whether such measurements can permit the prediction of: (a) whether a patient with CIS or RIS is likely to transit into definite MS, (b) whether a patient with MS is likely to show disease relapse in the near future, and/or (c) whether a patient with CIS, RIS or MS is likely to respond to/benefit from immune modulatory treatment.

While our data imply that there is indeed a correlation between blood cytokine responses and the clinical course of EAE, they leave open the question if and how the CNS cytokine response is linked to this correlation. It is tempting to speculate that the magnitude of the blood cytokine response defines the magnitude of the CNS cytokine response. On the one hand, it is conceivable that the more antigen-specific cells are present in the blood the more enter the CNS. On the other hand, it is equally possible that the more cells are present in the CNS, the less cells will be found in the blood. In addition, it needs to be defined if the magnitude of the IFN-γ/IL-17 response in the blood and/or the CNS itself is related to the degree of histopathology and if this finally defines the clinical outcome of the disease. Should we find a positive correlation between all four parameters - peripheral/CNS cytokine response, CNS pathology and clinical disease - our data would point towards the possibility of mirroring/predicting the degree of CNS pathology by simple measurements of blood responses and kinetics. It remains to be elucidated in the future whether our notion proves true.

5. Conclusion

What is to be concluded from the data presented here is that despite all the criticism about the model, EAE can be a valuable tool for studying MS. Different models can be used to selectively reflect different pathomechanisms of the disease. Most of the therapeutic options we rely on in the therapy of the disease today have been developed and/or validated in EAE and our data delineate that the model can also be used for working towards a better understanding of the interaction between the peripheral and CNS immune response and the subsequent damage to the target organ that finally defines the clinical outcome of the disease. The translation of our data into the clinical situation will show how accurate our interpretation of the disease processes presented here is.

6. Acknowledgement

We would like to thank Helena Batoulis, Traugott L. Gruppe, Lorenz C. Hundgeburth, Christian Kerkloh, Robert Pauly, Mascha S. Recks and Achim Schickel for scientific support and valuable discussions.

7. References

Ackerley, S., Grierson, A.J., Brownlees, J., Thornhill, P., Anderton, B.H., Leigh, P.N., Shaw, C.E. & Miller, C.C. (2000). Glutamate slows axonal transport of neurofilaments in transfected neurons, *J Cell Biol*, 150: 165-715

Aloisi, F. & Pujol-Borrell, R. (2006). Lymphoid neogenesis in chronic inflammatory diseases, *Nat Rev Immunol*, 6: 205-217

Armengol, M.P., Juan, M., Lucas-Martin, A., Fernandez-Figueras, M.T., Jaraquemada, D., Gallart, T. & Pujol-Borrell, R. (2001). Thyroid autoimmune disease: demonstration of thyroid antigen-specific B cells and recombination-activating gene expression in chemokine-containing active intrathyroidal germinal centers, *Am J Pathol*, 159: 861-873

Armengol, M.P., Cardoso-Schmidt, C.B., Fernández, M., Ferrer, X., Pujol-Borrell, R. & Juan, M. (2003). Chemokines determine local lymphangiogenesis and a reduction of circulating CXCR4+ and CCR7 B and T lymphocytes in thyroid autoimmune disease, *J Immunol*, 170: 6320-6328

Aziz, K.E., McCluskey, P.J. & Wakefield, D. (1997). Characterization of follicular dendritic cells in labial salivary glands of patients with primary Sjögren's syndrome: comparison with tonsillar lymphoid follicles, *Ann Rheum Dis*, 56: 140-143

Baddoura, F.K., Nasr, W., Wrobel, B., Li, Q., Ruddle, N.H. & Lakkis, F.G. (2005). Lymphoid neogenesis in murine cardiay allografts undergoing chronic rejection, *Am J Transplant* 2005, 5: 510-516

Bannerman, P.G., Hahn, A., Ramirez, S., Morley, M., Bönnemann, C., Yu, S., Zhang, G.X., Rostami, A. & Pleasure, D. (2005). Motor neuron pathology in experimental autoimmune encephalomyelitis: studies in THY1-YFP transgenic mice, *Brain*, 128: 1877-1886

Barone, F., Bombardieri, M., Manzo, A., Blades, M.C., Morgan, P.R., Challacombe, S.J., Valesini, G. & Pitzalis, C. (2008). Association of CXCL13 and CCL21 expression in salivary glands of patients with Sjögren's syndrome and MALT lymphoma: association with reactive and malignant areas of lymphoid organization, *J Immunol* 2008; 180: 5130-5140

Berger, T., Weerth, S., Kojima, K., Linington, C., Wekerle, H. & Lassmann, H. (1997). Experimental autoimmune encephalomyelitis: the antigen specificity of T lymphocytes determines the topography of lesions in the central and peripheral nervous system, *Lab Invest*, 76: 355-364.

Bernard, C.C. (1976). Experimental autoimmune encephalomyelitis in mice: genetic control of susceptibility, *J Immunogenet*, 3: 263-274

Bernard, C.C. & Carnegie, P.R. (1975). Experimental autoimmune encephalomyelitis in mice: immunologic response to mouse spinal cord and myelin basic proteins, *J Immunol*, 114: 1537-1540

Bo, L., Vedeler, C.A., Nyland, H.I., Trapp, B.D. & Mork, S.J. (2003). Subpial demyelination in the cerebral cortex of multiple sclerosis patients. *J Neuropathol Exp Neurol*, 62: 723-732

Burns, F.R., Li, X.B., Shen, N., Offner, H., Chou, Y.K., Vandenbark, A.A. & Heber-Katz, E. (1989). Both rat and mouse T cell receptors specific for the encephalitogenic determinant of myelin basic protein use similar V alpha and V beta chain genes even though the major histocompatibility complex and encephalitogenic determinants being recognized are different, *J Exp Med*, 169: 27-39

Cebecauer, M., Guillaume, P., Hozak, P., Mark, S., Everett, H., Schneider, P. & Luescher, I.F. (2005). Soluble MHC-peptide complexes induce rapid death of CD8+ CTL, *J Immunol*, 174: 6809-6819

Coronella, J.A., Spier, C., Welch, M., Trevor, K.T., Stopeck, A.T., Villar, H. & Hersh, E.M. (2002). Antigen-driven oligoclonal expansion of tumor-infiltrating B cells in infiltrating ductal carcinoma of the breast, *J Immunol*, 169: 1829-1836

Cross, A.H., Trotter, J.L. & Lyons, J. (2001) B cells and antibodies in CNS demyelinating disease, *J Neuroimmunol*, 112: 1-14

Derfuss, T., Parikh, H., Velhin, S., Braun, M., Mathey, E., Krumbholz, M., Kümpfel, T., Moldenhauer, A., Rader, C., Sonderegger, P., Pöllmann, W., Tiefenthaller, C., Bauer, J., Lassmann, H., Wekerle, H., Karagogeos, D., Hohlfeld, R., Linington, C. & Meinl, E. (2009). Contactin-2/TAG-1-directed autoimmunity is identified in multiple sclerosis patients and mediates gray matter pathology in animals, *Proc Natl Acad Sci USA*, 106: 8302-8307

de Rosbo, N.K., Kaye, J.F., Eisenstein, M., Mendel, I., Hoeftberger, R., Lassmann, H., Milo, R. & Ben Nun, A. (2004). The myelin-associated oligodendrocytic basic protein region MOBP15-36 encompasses the immunodominant major encephalitogenic epitope(s) for SJL/J mice and predicted epitope(s) for multiple sclerosis-associated HLA-DRB1*1501, *J Immunol*, 173: 1426-1435

Deteix, C., Attuil-Audenis, V., Duthey, A., Patey, N., McGregor, B., Dubois, V., Caligiuri, G., Graff-Dubois, S., Morelon, E. & Thaunat, O. (2001). Ingraft Th17 infiltrate promotes lymphoid neogenesis and hastens clinical chronic rejection, *J Immunol*, 184: 5344-5351

Dittmer, U. & Hasenkrug, K.J. (1999). Alternative interpretation of lymphocyte depletion studies using monoclonal antibodies in animals previously vaccinated with attenuated retroviral vaccines, *AIDS Res Hum Retroviruses*, 15: 785.

Drayton, D.L., Liao, S., Mounzer, R.H. & Ruddle, N.H. (2006). Lymphoid organ development: from ontogeny to neogenesis, *Nat Immunol*, 7: 344-353

Einstein, E.R., Robertson, D.M., Dicaprio, J.M. & Moore, W. (1962). The isolation from bovine spinal cord of a homogeneous protein with encephalitogenic activity, *J Neurochem*, 9: 353-361

Elliott, E.A., McFarland, H.I., Nye, S.H., Cofiell, R., Wilson, T.M., Wilkins, J.A., Squinto, J.P., Matis, L.A. & Mueller, J.P. (1996). Treatment of experimental encephalomyelitis with a novel chimeric fusion protein of myelin basic protein and proteolipid protein, *J Clin Invest*, 98: 1602-1612

Eugster, H.P., Frei, K., Bachmann, R., Bluethmann, H., Lassmann, H. & Fontana, A. (1999). Severity of symptoms and demyelination in MOG-induced EAE depends on TNFR1, *Eur J Immunol*, 29: 626-632

Fisher, E., Lee, J.C., Nakamura, K. & Rudick, R.A. (2008). Gray matter atrophy in multiple sclerosis: a longitudinal study, *Ann Neurol*, 64: 255-265

Forsthuber, T., Yip, H.C. & Lehmann, P.V. (1996). Induction of TH1 and TH2 immunity in neonatal mice, *Science*, 271: 1728-1730

Fritz, R.B., Chou, C.H. & McFarlin, D.E. (1983). Induction of experimental allergic encephalomyelitis in PL/J and (SJL/J x PL/J)F1 mice by myelin basic protein and its peptides: localization of a second encephalitogenic determinant, *J Immunol*, 130: 191-194

Fritz, R.B., Skeen, M.J., Chou, C.H., Garcia & Egorov, I.K. (1985). Major histocompatibility complex-linked control of the murine immune response to myelin basic protein, *J Immunol*, 134: 2328-2332

Fritz, R.B. & Zhao, M.L. (1996). Active and passive experimental autoimmune encephalomyelitis in strain 129/J (H-2b) mice, *J Neurosci Res*, 45: 471-474

Galboiz, Y. & Miller, A. (2002). Immunological indicators of disease activity and prognosis in multiple sclerosis, *Curr Opin Neurol*, 15: 233-237

Gasser, D.L., Goldner-Sauve, A. & Hickey, W.F. (1990). Genetic control of resistance to clinical EAE accompanied by histological symptoms, *Immunogenetics*, 31: 377-382

Gold, R., Linington, C. & Lassmann, H. (2006). Understanding pathogenesis and therapy of multiple sclerosis via animal models: 70 years of merits and culprits in experimental autoimmune encephalomyelitis research, *Brain*, 129: 1953-1971

Goverman, J. & Brabb, T. (1996). Rodent models of experimental allergic encephalomyelitis applied to the study of multiple sclerosis, *Lab Anim Sci*, 46: 482-492

Hafler, D.A., Slavik, J.M., Anderson, D.E., O'Connor, K.C., De Jager, P. & Baecher-Allan, C. (2005). Multiple sclerosis, *Immunol Rev*, 204: 208-231

Hardin-Pouzet, H., Krakowski, M., Bourbonniere, L., Didier-Bazes, M., Tran, E. & Owens, T. (1997). Glutamate metabolism is down-regulated in astrocytes during experimental allergic encephalomyelitis, *Glia*, 20: 79-85

Hashim, G.A., Sharpe, R.D., Carvalho, E. & Stevens, L.E. (1975). The development of experimental allergic encephalomyelitis with immunizing doses of myelin basic protein, *Proc Soc Exp Biol Med*, 149: 646-651

Hemmer, B., Cepok, S., Nessler, S. & Sommer, N. (2002). Pathogenesis of multiple sclerosis: an update on immunology, *Curr Opin Neurol*, 15: 227-231

Hofstetter, H.H., Shive, C.L. & Forsthuber, T.G. (2002). Pertussis toxin modulates the immune response to neuroantigens injected in incomplete Freund's adjuvant: induction of Th1 cells and experimental autoimmune encephalomyelitis in the presence of high frequencies of Th2 cells, *J Immunol*, 169: 117-125

Holz, A., Bielekova, B., Martin, R. & Oldstone, M.B. (2000). Myelin-associated oligodendrocytic basic protein: identification of an encephalitogenic epitope and association with multiple sclerosis, *J Immunol*, 164: 1103-1109

Hoppenbrouwers, I..A. & Hintzen, R.Q. (2010). Genetics of multiple sclerosis. *Biochim Biophys Acta*, 1812: 194-201

Inglese, M., Oesingmann, N., Casaccia, P. & Fleysher, L. (2011). Progressive multiple sclerosis and gray matter pathology: an MRI perspective, *Mt Sinai J Med*, 78: 258-267

Jansson, L., Diener, P., Engstrom, A., Olsson, T. & Holmdahl, R. (1995). Spreading of the immune response to different myelin basic protein peptides in chronic experimental autoimmune encephalomyelitis in B10.RIII mice, *Eur J Immunol*, 25: 2195-2200

Jordan, E.K., McFarland, H.I., Lewis, B.K., Tresser, N., Gates, M.A., Johnson, M., Lenardo, M., Matis, L.A., McFarland, H.F. & Frank, J.A. (1999). Serial MR imaging of experimental autoimmune encephalomyelitis induced by human white matter or by chimeric myelin-basic and proteolipid protein in the common marmoset, *Am J Neuroradiol*, 20: 965-976

Kasper, L.H., Haque, A &, Haque, S. (2007). Regulatory mechanisms of the immune system in multiple sclerosis. T regulatory cells: turned on to turn off, *J Neurol*, 254 [Suppl1]: I/10-I/14

Kidd, D., Barkhof, F., McConnell, R., Algra, P.R., Allen, I.V. & Revesz, T. (1999). Cortical lesions in multiple sclerosis, *Brain*, 122: 17-26

Kipp, M., Clarner, T., Dang, J., Copray, S. & Beyer, C. (2009). The cuprizone animal model: new insights into an old story, *Acta Neuropathol*, 118: 723-736

Klein, L., Klugmann, M., Nave, K.A., Tuohy, V.K. & Kyewski, B. (2000). Shaping of the autoreactive T-cell repertoire by a splice variant of self protein expressed in thymic epithelial cells, *Nat Med*, 6: 56-61

Kojima, K., Wekerle, H., Lassmann, H., Berger, T. & Linington, C. (1997). Induction of experimental autoimmune encephalomyelitis by CD4+ T cells specific for an astrocyte protein, S100 beta, *J Neural Transm Suppl*, 49: 43-51

Kooi, E., Geurts, J.J.G., van Horssen, J., Bo, L. & van der Valk, P. (2009). Meningeal inflammation is not associated with cortical demyelination in chronic multiple sclerosis, *J Neuropathol Exp Neurol*, 68: 1021-1028

Kuchroo, V.K., Sobel, R.A., Laning, J.C., Martin, C.A., Greenfield, E., Dorf, M.E. & Lees, M.B. (1992). Experimental allergic encephalomyelitis mediated by cloned T cells specific for a synthetic peptide of myelin proteolipid protein. Fine specificity and T cell receptor V beta usage, *J Immunol*, 148: 3776-3782

Kuerten, S., Lichtenegger, F.S., Faas, S., Angelov, D.N., Tary-Lehmann, M. & Lehmann, P.V. (2006). MBP-PLP fusion protein-induced EAE in C57BL/6 mice, *J Neuroimmunol*, 177: 99-111

Kuerten, S., Kostova-Bales, D.A., Frenzel, L.P., Tigno, J.T., Tary-Lehmann, M., Angelov, D.N. & Lehmann, P.V. (2007). MP4- and MOG:35-55-induced EAE in C57BL/6 mice differentially targets brain, spinal cord and cerebellum, *J Neuroimmunol*, 189: 31-40

Kuerten, S., Javeri, S., Tary-Lehmann, M., Lehmann, P.V. & Angelov, D.N. (2008). Fundamental differences in the dynamics of CNS lesion development and composition in MP4- and MOG peptide 35-55-induced experimental autoimmune encephalomyelitis, *Clin Immunol*, 129: 256-267

Kuerten, S., Rodi, M., Javeri, S., Gruppe, T.L., Tary-Lehmann, M., Lehmann, P.V. & Addicks, K. (2009). Delineating the impact of neuroantigen vs genetic diversity on MP4-induced EAE of C57BL/6 and B6.129 mice, *APMIS*, 117: 923-935

Kuerten, S., Rottlaender, A., Rodi, M., Velasco, V.B. Jr., Schroeter, M., Kaiser, C., Addicks, K., Tary-Lehmann, M. & Lehmann, P.V. (2010). The clinical course of EAE is reflected by the dynamics of the neuroantigen-specific T cell compartment in the blood, *Clin Immunol*, 137: 422-432

Kuerten, S., Pauly, R., Rottlaender, A., Rodi, M., Gruppe, T.L., Addicks, K., Tary-Lehmann, M. & Lehmann, P.V. (2011a). Myelin-reactive antibodies mediate the pathology of MBP-PLP fusion protein MP4-induced EAE, *Clin Immunol*, 140: 54-62

Kuerten, S., Gruppe, T.L., Laurentius, L.M., Kirch, C., Lehmann, P.V. & Addicks, K. (2011b). Differential patterns of spinal cord pathology induced by MP4, MOG peptide 35-55, and PLP peptide 178-191 in C57BL/6 mice, *APMIS*, 119: 336-346

Kuerten, S., Kerkloh, C., Schickel, A., Addicks, K., Lehmann, P.V. & Ruddle, N.H. (2011c). Tertiary lymphoid organs in MP4-induced EAE facilitate determinant spreading to CNS antigens, manuscript in preparation

Hundgeburth, L.C., Pauly, R., Lehmann, P.V. & Kuerten, S. (2011d). CNS pathology mediated by MP4-specific antibodies is complement-dependent, manuscript in preparation

Gruppe, T.L., Recks, M.S., Addicks, K. & Kuerten, S. (2011). Differential contribution of spinal cord white and gray matter pathology to experimental autoimmune encephalomyelitis, manuscript in preparation

Kumar, V. & Sercarz, E. (1994). Holes in the T cell repertoire to myelin basic protein owing to the absence of the D beta 2-J beta 2 gene cluster: implications for T cell receptor recognition and autoimmunity, *J Exp Med*, 179: 1637-1643

Kutzelnigg, A., Lucchinetti, C.F., Stadelmann, C., Brück, W., Rauschka, H., Bergmann, M., Schmidbauer, M., Parisi, J.E. & Lassmann, H. (2005). Cortical demyelination and diffuse white matter injury in multiple sclerosis, *Brain*, 128: 2705-2712

Kutzelnigg, A., Faber-Rod, J.C., Bauer, J., Lucchinetti, C.F., Sorensen, P.S., Laursen, H., Stadelmann, C., Brück, W., Rauschka, H., Schmidbauer, M. & Lassmann, H. (2007). Widespread demyelination in the cerebellar cortex in multiple sclerosis, *Brain Pathol*, 17: 38-44

Lebar, R., Lubetzki, C., Vincent, C., Lombrail, P. & Boutry, J.M. (1986). The M2 autoantigen of central nervous system myelin, a glycoprotein present in oligodendrocyte membrane, *Clin Exp Immunol*, 66: 423-434

Lehmann, P.V., Forsthuber, T., Miller, A. & Sercarz, E.E. (1992). Spreading of T-cell autoimmunity to cryptic determinants of an autoantigen, *Nature*, 358: 155-157

Lucchinetti, C., Brück, W., Parisi, J., Scheithauer, B., Rodriguez, M. & Lassmann, H. (2000). Heterogeneity of multiple sclerosis lesions: implications for the pathogenesis of demyelination, *Ann Neurol*, 47: 707-717

Lyons, J.A., Ramsbottom, M.J. & Cross, A.H. (2002). Critical role of antigen-specific antibody in experimental autoimmune encephalomyelitis induced by recombinant myelin oligodendrocyte glycoprotein, *Eur J Immunol*, 32: 1905-1913

Magliozzi, R., Howell, O., Vora, A., Serafini, B., Nicholas, R., Puopolo, M., Reynolds, R. & Aloisi, F. (2007). Meningeal B-cell follicles in secondary progressive multiple sclerosis associate with early onset of disease and severe cortical pathology, *Brain*, 130: 1089-1094

Marta, C.B., Oliver, A.R., Sweet, R.A., Pfeiffer, S.E., Ruddle, N.H. (2005). Pathogenic myelin oligodendrocyte glycoprotein antibodies recognize glycosylated epitopes and perturb oligodendrocyte physiology, *Proc Natl Acad Sci USA*, 102: 13992-13997

Martenson, R.E., Deibler, G.E. & Kies, M.W. (1970). Myelin basic proteins of the rat central nervous system. Purification, encephalitogenic properties, and amino acid compositions, *Biochem Biophys Acta*, 200: 353-362

Matute, C., Alberdi, E., Domercq, M., Perez-Cerda, F., Perez-Samartin, A. & Sanchez-Gomez, M.V. (2001). The link between excitotoxic oligodendroglial death and demyelinating diseases, *Trends Neurosci*, 24: 224-230

Mazzucchelli, L., Blaser, A., Kappeler, A., Schärli, P., Laissue, J.A., Baggiolini, M. & Uguccioni, M. (1999). BCA-1 is highly expressed in Helicobacter pylori-induced mucosa-associated lymphoid tissue and gastric lymphoma, *J Clin Invest*, 104: R49-54

McRae, B.L., Vanderlugt, C.L., Dal Canto, M.C. & Miller, S.D. (1995). Functional evidence for epitope spreading in the relapsing pathology of experimental autoimmune encephalomyelitis, *J Exp Med*, 182: 75-85

Mendel, I., de Rosbo, N.K. & Ben Nun, A. (1995). A myelin oligodendrocyte glycoprotein peptide induces typical chronic experimental autoimmune encephalomyelitis in H-2b mice: fine specificity and T cell receptor V beta expression of encephalitogenic T cells, *Eur J Immunol*, 25: 1951-1959

Mendel, I., de Rosbo, N.K., & Ben Nun, A. (1996). Delineation of the minimal encephalitogenic epitope within the immunodominant region of myelin oligodendrocyte glycoprotein: diverse V beta gene usage by T cells recognizing the core epitope encephalitogenic for T cell receptor V beta b and T cell receptor V beta a H-2b mice, *Eur J Immunol*, 26: 2470-2479

Mor, F., Quintana, F., Mimran, A. & Cohen, I.R. (2003). Autoimmune encephalomyelitis and uveitis induced by T cell immunity to self beta-synuclein, *J Immunol*, 170: 628-634

Morris-Downes, M.M., McCormack, K., Baker, D., Sivaprasad, D., Natkunarajah, J. & Amor, S. (2002). Encephalitogenic and immunogenic potential of myelin-associated glycoprotein (MAG), oligodendrocyte-specific glycoprotein (OSP) and 2',3'-cyclic nucleotide 3'-phosphodiesterase (CNPase) in ABH and SJL mice, *J Neuroimmunol*, 122: 20-33

Murakami, J., Shimizu, Y., Kashii, Y., Kato, T., Minemura, M., Okada, K., Nambu, S., Takahara, T., Higuchi, K., Maeda, Y., Kumada, T. &Watanabe, A. (1999). Functional

B-cell response in intrahepatic lymphoid follicles in chronic hepatitis C, *Hepatology*, 30: 143-150

Nzula, S., Going, J.J. & Stott, D.I. (2003). Antigen-driven clonal proliferation, somatic hypermutation, and selection of B lymphocytes infiltrating human ductal breast carcinomas, *Cancer Res*, 63: 3275-3280

Oliver, A.R., Lyon, G.M. & Ruddle, N.H. (2003). Rat and human myelin oligodendrocyte glycoproteins induce experimental autoimmune encephalomyelitis by different mechanisms in C57BL/6 mice, *J Immunol*, 171: 462-468

Peferoen, L.A.N., Lamers, F., Lodder, L.N.R., Gerritsen, W.H., Huitinga, I., Melief, J., Giovannoni, G., Meier, U., Hintzen, R.Q., Verjans, G.M., van Nierop, G.P., Vos, W., Peferoen-Baert, R.M., Middeldorp, J.M., van der Valk, P. & Amor, S. (2010). Epstein Barr virus is not a characteristic feature in the central nervous system in established multiple sclerosis, *Brain*, 133: e137

Pöllinger, B., Krishnamoorthy, G., Berer, K., Lassmann, H., Bösl, M.R., Dunn, R., Dominques, H.S., Holz, A., Kurschus, F.S. & Wekerle, H. (2009). Spontaneous relapsing-remitting EAE in the SJL/J mouse: MOG-reactive transgenic T cells recruit endogenous MOG-specific B cells, *J Exp Med*, 206: 1303-1316

Radu, C.G., Anderton, S.M., Firan, M., Wraith, D.C. & Ward, E.S. (2000). Detection of autoreactive T cells in H-2u mice using peptide-MHC multimers, *Int Immunol*, 12: 1553-1560

Reindl, M., Khalil, M. & Berger, T. (2006). Antibodies as biological markers for pathophysiological processes in MS, *J Neuroimmunol*, 180: 50-62

Rinaldi, L. & Gallo, P. (2005). Immunological markers in multiple sclerosis: tackling the missing elements, *Neurol Sci*, 26: S15-S17

Rudick, R.A. & Trapp, B.D. (2009). Gray-matter injury in multiple sclerosis, *N Engl J Med*, 361: 1505-150

Sadler, R.H., Sommer, M.A., Forno, L.S., Smith, M.E. (1991). Induction of anti-myelin antibodies in EAE and their possible role in demyelination, *J Neurosci Res*, 30: 616-624

Salomonsson, S., Jonsson, M.V., Skarstein, K., Brokstad, K.A., Hjelmström, P., Wahren-Herlenius, M. & Jonsson, R. (2003). Cellular basis of ectopic germinal center formation and autoantibody production in the target organ of patients with Sjögren's syndrome, *Arthritis Rheum*, 48: 3187-3201

Schmidt, S. (1999). Candidate autoantigens in multiple sclerosis, *Mult Scler*, 5: 147-160

Serafini, B., Rosicarelli, B., Magliozzi, R., Stigliano, A. & Aloisi, F. (2004). Detection of ectopic B-cell follicles with germinal centers in meninges of patients with secondary progressive multiple sclerosis, *Brain Pathol*, 14: 164-174

Serafini, B., Severa, M., Columba-Cabezas, S., Rosicarelli, B., Veroni, C., Chiappetta, G., Magliozzi, R., Reynolds, R., Coccia, E.M. & Aloisi, F. (2010). Epstein-Barr virus latent infection and BAFF expression in B cells in the multiple sclerosis brain: implications for viral persistence and intrathecal B-cell activation, *J Neuropathol Exp Neurol*, 69: 677-693

Shi, K., Hayashida, K., Kaneko, M., Hashimoto, J., Tomita, T., Lipsky, P.E., Yoshikawa, H. & Ochi, T. (2001). Lymphoid chemokine B cell-attracting chemokine-1 (CXCL13) is expressed in germinal center of ectopic lymphoid follicles within the synovium of chronic arthritis patients, *J Immunol*, 166: 650-655

Sicotte, N.L. (2011). Neuroimaging in multiple sclerosis: neurotherapeutic implications, *Neurotherapeutics*, 8: 54-62

Silvera, P., Wade-Evans, A., Rud, E., Hull, R., Silvera, K., Sangster, R., Almond, N. & Stott, J. (2001). Mechanisms of protection induced by live attenuated simian immunodeficiency virus: III. Viral interference and the role of CD8+ T-cells and beta-chemokines in the inhibition of virus infection of PBMCs in vitro, *J Med Primatol*, 30: 1-13

Sims, G.P., Shiono, H., Willcox, N. & Stott, D.I. (2001). Somatic hypermutation and selection of B cells in thymic germinal centers responding to acetylcholine receptor in myasthenia gravis, *J Immunol*, 167:1935-1944

Skundric, D.S., Huston, K., Shaw, M., Tse, H.Y. & Raine, C.S. (1994). Experimental allergic encephalomyelitis. T cell trafficking to the central nervous system in a resistant Thy-1 congenic mouse strain, *Lab Invest*, 71: 671-679

Smith, T., Groom, A., Zhu, B. & Turski, L. (2000). Autoimmune encephalomyelitis ameliorated by AMPA antagonists, *Nat Med*, 6: 62-66

Stefferl, A., Brehm, U. & Linington C. (2000). The myelin oligodendrocyte glycoprotein (MOG): a model for antibody-mediated demyelination in experimental autoimmune encephalomyelitis and multiple sclerosis, *J Neural Transm Suppl*, 58: 123-133

Steinman, L. (1999). Assessment of animal models for MS and demyelinating disease in the design of rational therapy, *Neuron*, 24: 511-514

Steinman, L. & Zamvil, S.S. (2006). How to successfully apply animal studies in experimental autoimmune encephalomyelitis to research on multiple sclerosis, *Ann Neurol*, 60: 12-21

Stott, D.I., Hiepe, F., Hummel, M., Steinhauser, G. & Berek, C. (1998). Antigen-driven clonal proliferation of B cells within the target tissue of an autoimmune disease. The salivary glands of patients with Sjögren's syndrome, *J Clin Invest*, 102: 938-946

Takemura, S., Klimiuk, P.A., Braun, A., Goronzy, J.J. & Weyand, C.M. (2001). T cell activation in rheumatoid synovium is B cell dependent, *J Immunol* 167, 4710-4718

Targoni, O.S., Baus, J., Hofstetter, H.H., Hesse, M.D., Karulin, A.Y., Boehm, B.O., Forsthuber, T.G. & Lehmann, P.V. (2001) Frequencies of neuroantigen-specific T cells in the central nervous system versus the immune periphery during the course of experimental allergic encephalomyelitis, *J Immunol*, 166: 4757-4764

Tompkins, S.M., Padilla, J., Dal Canto, M.C., Ting, J.P., Van Kaer, L. & Miller, S.D. (2002). De novo central nervous system processing of myelin antigen is required for the initiation of experimental autoimmune encephalomyelitis, *J Immunol*, 168: 4173-4183

Trapp, B.D., Peterson, J., Ransohoff, R.M., Rudick, R., Mörk, S. & Bö, L (1996). Axonal transection in the lesions of multiple sclerosis, *N Engl J Med*, 338: 278-285

Tsunoda, I. &Fujinami, R.S. (2010). Neuropathogenesis of Theiler's murine encephalomyelitis virus infection, an animal model for multiple sclerosis, *J Neuroimmune Pharmacol*, 5: 355-369

Tuohy, V.K. (1993). Peptide determinants of myelin proteolipid protein (PLP) in autoimmune demyelinating disease : a review, *Neurochem Res*, 19: 935-944

Tuohy, V.K., Yu, M., Yin, L., Kawczak, J.A., Johnson, J.M., Mathisen, P.M., Weinstock-Guttman, B. & Kinkel, R.P. (1998). The epitope spreading cascade during progression of experimental autoimmune encephalomyelitis and multiple sclerosis, *Immunol Rev*, 164: 93-100

Tuohy, V.K., Yu, M., Yin, L., Kawczak, J.A. & Kinkel, P.R. (1999). Regression and spreading of self-recognition during the development of autoimmune demyelinating disease, *J Autoimmun*, 13: 11-20

Urban, J.L., Kumar, V., Kono, D.H., Gomez, C., Horvath, S.J., Clayton, J., Ando, D.G., Sercarz, E.E. & Hood, L. (1988). Restricted use of T cell receptor V genes in murine autoimmune encephalomyelitis raises possibilities for antibody therapy, *Cell*, 54: 577-592

Van Epps, H.L. (2005). Thomas Rivers and the EAE model, *J Exp Med*, 202: 4

Vogt, J., Paul, F., Aktas, O., Müller-Wielsch, K., Dörr, J., Dörr, S., Bharathi, B.S., Glumm, R., Schmitz, C., Steinbusch, H., Raine, C.S., Tsokos, M., Nitsch, R. & Zipp, F. (2009). Lower motor neuron loss in multiple sclerosis and experimental autoimmune encephalomyelitis, *Ann Neurol*, 66: 310-322

Waxman, F.J., Fritz, R.B. & Hinrichs, D.J. (1980). The presence of specific antigen-reactive cells during the induction, recovery, and resistance phases of experimental allergic encephalomyelitis, *Cell Immunol*, 49: 34-42

Weerth, S., Berger, T., Lassmann, H. & Linington, C. (1999). Encephalitogenic and neuritogenic T cell responses to the myelin-associated glycoprotein (MAG) in the Lewis rat, *J Neuroimmunol*, 95: 157-164

Wegner, C., Matthews, P.M. (2003). A new view of the cortex, new insights into multiple sclerosis, *Brain*, 126: 1719-1721

Werner, P., Pitt, D. & Raine, C.S. (2011). Multiple sclerosis: altered glutamate homeostasis in lesions correlates with oligodendrocyte and axonal damage, *Ann Neurol*, 50: 169-180

Willis, S.N., Stadelmann, C., Rodig, S.J., Caron, T., Gattenloehner, S., Mallozzi, S.S., Roughan, J.E., Almendinger, S.E., Blewett, M.M., Brück, W., Hafler, D.A., O'Connor, K.C. (2009). Epstein-Barr virus infection is not a characteristic feature of multiple sclerosis brain, *Brain*, 132: 3318-3328

Yip, H.C., Karulin, A.Y., Tary-Lehmann, M., Hesse, M.D., Radeke, H., Heeger, P.S., Trezza, R.P., Heinzel, F.P., Forsthuber, T. & Lehmann, P.V. (1999). Adjuvant-guided type-1 and type-2 immunity: infectious/noninfectious dichotomy defines the class of response, *J Immunol*, 162: 3942-3949

Yu, M., Johnson, J.M., Tuohy, V.K. (1996). A predictable sequential determinant spreading cascade invariably accompanies progression of experimental autoimmune encephalomyelitis: a basis for peptide-specific therapy after onset of clinical disease, *J Exp Med*, 183: 1777-1788

Zamvil, S.S., Mitchell, D.J., Lee, N.E., Moore, A.C., Waldor, M.K., Sakai, K., Rothbard, J.B., McDevitt, H.O., Steinman, L. & Acha-Orbea, H. (1988). Predominant expression of a T cell receptor V beta gene subfamily in autoimmune encephalomyelitis, J Exp Med, 167: 1586-1596

Zhu, B., Luo, L., Moore, G.R., Paty, D.W., Cynader, M.S. (2003). Dendritic and synaptic pathology in experimental autoimmune encephalomyelitis, *Am J Pathol*, 162: 1639-1650

3

Assessment of Neuroinflammation in Transferred EAE Via a Translocator Protein Ligand

F. Mattner et al.*
[1]*Life Sciences Division, Australian Nuclear Science and Technology Organisation*
Australia

1. Introduction

Neuroinflammation is involved in the pathogenesis and progression of neurological disorders such as Alzheimer's disease and multiple sclerosis (MS) (Doorduin et al., 2008). MS has been considered a T cell-mediated autoimmune disorder of the central nervous system (CNS), characterized by inflammatory cell infiltration and myelin destruction (Hauser et al., 1986) and focal demyelinated lesions in the white matter are the traditional hallmarks of MS. However more recent evidence suggests more widespread damage to the brain and spinal cord, to areas of white matter distant from the inflammatory lesions and demyelination of deep and cortical grey matter (McFarland & Martin, 2007). Experimental autoimmune encephalomyelitis (EAE) is an extensively used model of T-cell mediated CNS inflammation; modelling disease processes involved in MS. EAE can be induced in several species by immunization with myelin antigens or via adoptive transfer of myelin-reactive T cells. The models of EAE in rodents [actively induced and transferred] provide information about different phases [inflammation, demyelination and remyelination] and types [monophasic, chronic-relapsing and chronic-progressive] of the human disease multiple sclerosis and a vast amount of clinical and histopathologic data has been accumulated through the decades. A key aim of current investigations is developing the ability to recognise the early symptoms of the disease and to follow its course and response to treatment.

Molecular imaging is a rapidly evolving field of research that involves the evaluation of biochemical and physiological processes utilising specific, radioactive, fluorescent and magnetic resonance imaging probes. However, it is positron emission tomography (PET) and single photon emission computer tomography (SPECT) which, due to their exquisite sensitivity involving specifically designed radiolabelled molecules, that is leading the way in molecular imaging and has greatly enabled the non-invasive "visualisation" of many diseases in both animal models and humans. Furthermore, PET and SPECT molecular

* M. Staykova[2], P. Callaghan[1], P. Berghofer[1], P. Ballantyne[1], M.C. Gregoire[1], S. Fordham[2], T. Pham[1], G. Rahardjo[1], T. Jackson[1], D. Linares[2] and A. Katsifis[1]
[1]*Life Sciences Division, Australian Nuclear Science and Technology Organisation, Australia*
[2]*Neurosciences Research Unit, The Canberra Hospita, Australia*

imaging are providing invaluable imaging data based on a biochemical-molecular biology interaction rather than from the traditional anatomical view. Increasingly, PET and SPECT radiotracers have been exploited to study or identify molecular biomarkers of disease, monitor disease progression, determining the effects of a drug on a particular pathology and assess the pharmacokinetic behaviour of pharmaceuticals *in vivo*. Significantly, these new imaging systems provide investigators with an unprecedented ability to examine and measure *in vivo* biological and pharmacological processes over time in the same animals thus reducing experimental variability, time and costs. Molecular imaging based on the radiotracer principle allows chemical processes ranging from cellular events, to cellular communication and interaction in their environment, to the organisation and function of complete tissue and organs to be studied in real time without perturbation. One of the key benefits of molecular imaging is a technique that allows longitudinal studies vital for monitoring intra-individual progression in disease, or regression with supplementary pharmacotherapies. This is key in animal models of diseases such as MS, where there is significant intra-individual variability in the disease course and severity.

Recent investigations have proposed the translocator protein (TSPO; 18 kDa), also known as the peripheral benzodiazepine receptor (PBR), as a molecular target for imaging neuroinflammation (Chen & Guilarte, 2008; Doorduin et al., 2008; Papadopoulos et al., 2006). TSPO (18 kDa) is a multimeric protein consisting of five transmembrane helices, which, in association with a 32 kDa subunit that functions as a voltage dependent anion channel and a 30 kDa subunit that functions as an adenine nucleotide carrier forms part of a hetero-oligomeric complex (McEnery et al., 1992) responsible for cholesterol, heme and calcium transport in specific tissue. TSPO is primarily located on the outer mitochondrial membrane and is predominantly expressed in visceral organs (kidney, heart) and the steroid hormone producing cells of the adrenal cortex, testis and ovaries. In the central nervous system (CNS), TSPO is sparsely expressed under normal physiological conditions, however its expression is significantly upregulated following CNS injury (Chen et al., 2004; Papadopoulos et al., 1997; Venneti et al., 2006; Venneti, et al., 2008).

Several studies have identified activated glial cells as the cells responsible for TSPO upregulation in inflamed brain tissue, both in humans and in experimental models (Mattner et al., 2011; Myers et al., 1991a; Stephenson et al., 1995; Vowinckel et al., 1997) and the TSPO ligand [^{11}C]-PK11195 was one of the first PET ligands used for imaging activated microglia in various neurodegenerative diseases (Venneti et al., 2006). Although [^{11}C]-(R)-PK11195 is widely used for imaging of microglia, its considerable high plasma protein binding, high levels of nonspecific binding, relatively poor blood–brain barrier permeability and short half-life, limits its use in brain imaging (Chauveau et al., 2008). Recently, alternative PET radioligands for TSPO including the phenoxyarylacetamide derivative [^{11}C]-DAA1106 and its analogues (Gulyas et al., 2009; Takano et al., 2010; Venneti et al., 2008), the imidazopyridines (PBR111) and its analogues (Boutin et al., 2007a; Fookes et al., 2008) and the pyrazolo[1,5-a]pyrimidine derivatives [^{18}F]-DPA-714 and [^{11}C]-DPA-713 (Boutin et al., 2007b; James et al., 2008) have been investigated.

In addition to imaging with PET, recent advances in new generation of hybrid SPECT imaging systems enabling increased resolution and morphological documentation with associated computed tomography have been made for use clinically and preclinically. These advances have created a need and an opportunity for SPECT tracers; particularly those incorporating the longer lived radiotracer iodine-123 (t ½ = 13.2 h), to facilitate extended longitudinal imaging studies.

In this study the recently developed high-affinity TSPO, SPECT ligand, 6-chloro-2-(4'-iodophenyl)-3-(N,N-diethyl)-imidazo[1,2-a]pyridine-3-acetamide or CLINDE , was used to explore the expression of activated glia in a model of transferred EAE (tEAE). [^{123}I]-CLINDE has demonstrated its potency and specificity for TSPO binding, its ability to penetrate the blood-brain barrier and suitable pharmacokinetics for SPECT imaging studies (Mattner et al., 2008). It has also been shown that [^{123}I]-CLINDE was able to detect *in vivo* inflammatory processes characterized by increased density of TSPO in several animal models (Arlicot et al., 2008; Arlicot et al., 2010; Mattner et al., 2005; Mattner et al., 2011; Song et al., 2010), thus representing a promising SPECT radiotracer for imaging neuroinflammation. The present study aimed to investigate the effectiveness of [^{123}I]-CLINDE to detect and quantify the activated glia and consequently correlate the intensity of TSPO upregulation with the severity of disease in a model of tEAE.

2. *In vivo* distribution and *in vitro* binding of TSPO - correlation with upregulation in a model of tEAE

2.1 *In vivo* evaluation

The effectiveness of [^{123}I]-CLINDE to detect and quantify activated glia and correlate TSPO upregulation to the severity of neuroinflammation was assessed on a Lewis rat model of tEAE (Willenborg et al., 1986). The intravenous injection of myelin basic protein (MBP)-specific T lymphoblasts results in a single disease episode. The cells should be used when encephalitogenic, i.e. in the first three days after MBP re-stimulation of spleen or lymph node cells (from MBP- complete Freund's adjuvant (CFA) primed Lewis rats) or established MBP-specific $CD4^+$ IFNγ producing T line cells (De Mestre et al., 2007). A huge advantage of this model of neuroinflammation is its uniform time course.

Male Lewis rats (Animal Resource Centre, Australia) were maintained and monitored according to Australian laws governing animal experimentation. The rats were immunised with emulsion of bovine MBP and CFA (4 mg/ml Mycobacterium butyricum). On day 10 the popliteal and inguinal lymph nodes were removed and single cell suspensions were incubated with the antigens MBP or PPD (50 µg/ml) for 3 days. MBP and PPD lymphoblasts were isolated on a density gradient (d=1.077), propagated in IL-2 containing medium for 25 h and $3x10^7$ cells were injected via the lateral tail vein into naive Lewis rats. The rats were examined daily and a clinical score assigned according to the accepted scale: 0, asymptomatic; 1, flaccid distal half of the tail; 2, entire tail flaccid; 3, ataxia (difficulty righting). Half values were given when assessment fell between two scores.

The radiotracers [^{123}I]-CLINDE and [^{125}I]-CLINDE were synthesized as previously described (Katsifis et al., 2000) using an improved method, giving rise to high purity and high specific activity product. Briefly the tributyltin precursor (50-100 µg) in acetic acid (200 µL) was treated with a solution of either no carrier added $Na^{123}I$ in 0.02 M NaOH (Australian Radioisotopes and Industrials, Sydney, Australia) or $Na^{125}I$ (GE-Healthcare) followed by peracetic acid (1-3%, 100 µL). After 5 min the reaction was quenched (sodium bisulphite, 200 µL, 50 mg/mL), neutralised (sodium bicarbonate, 200 µl, 50 mg/mL) and injected onto a semipreparative C-18 RP-HPLC column. The purification and isolation of [123/125I]-CLINDE was carried out by C-18 RP-HPLC using a mixture of acetonitrile/0.1 M ammonium acetate 55:45 at a flow rate of 4 mL/min. Under these conditions, the radiotracer eluted at 25 min. The eluted product was evaporated to dryness and reconstituted in saline (0.9%) for *in vivo* pharmacological studies. For *in vitro* assays, unlabelled CLINDE was added to the [125I]-CLINDE to achieve a specific activity of 3.7 GBq/μmol and reconstituted in ethanol. The specific activity of [123I]-CLINDE was assumed to be greater than 185 GBq/μmol based on the limit of detection of the UV in the HPLC system used. The specific activity of [125I]-CLINDE was measured as 80 GBq/μmol, close to the theoretical specific activity of the [125I]-iodine.

In order to assess the distribution of the [123I]-CLINDE *in vivo*, rats showing different tEAE clinical scores were used: [0 (pre-clinical, *n* = 4), 2 (*n* = 4), 3 (*n* = 4) and 0 (after recovery, *n* = 3)]. Rats were given PPD-lymphoblasts (n = 4) and control naive rats (*n* = 4) served as controls. The animals were injected via the tail vein with 0.70 MBq of [123I]-CLINDE in saline. Tissue samples were taken 3 h later, the radioactivity was measured with an automated gamma counter and the percent injected dose (%ID/g) was calculated by comparison with samples of standard dilutions of the initial dose.

Statistically significant increase in [123I]-CLINDE uptake was measured in brains and spinal cords of Lewis rats given MBP-specific T lymphoblasts that developed EAE with clinical scores of 2 and 3 (Figure 1). In the brain, the medulla oblongata, medulla pons, cerebellum, diencephalon, hypothalamus, hippocampus, frontal and posterior cortex were affected and the increase in [123I]-CLINDE uptake was in the range of 1.5 - 3.8 times.

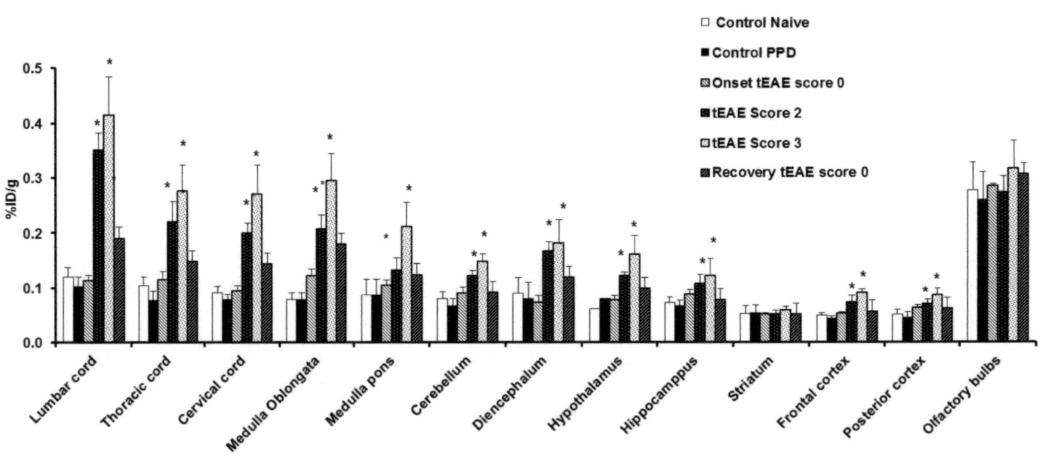

Fig. 1. Uptake of [123I]-CLINDE in the CNS. Data are represented as mean ± SD of wet tissue.* $p<0.05$ - One way ANOVA with Tukeys post-hoc test.

In the spinal cord the uptake of activity reflected the ascending nature of the inflammatory process: lumbar spinal cord > thoracic spinal cord > cervical spinal cord. The positive correlation between the ligand uptake and the disease severity is shown also in Figure 2.

Importantly, the radiotracer uptake on day 4 did not differ significantly from the one in the naive controls if the disease irrelevant PPD-specific T lymphoblasts or the disease relevant

MBP-specific T lymphoblasts were injected (Figure 1). Also, all animals showed the typical TSPO ligand biodistribution in the visceral organs and EAE severity had no influence on ligand uptake in the visceral organs (Table 1).

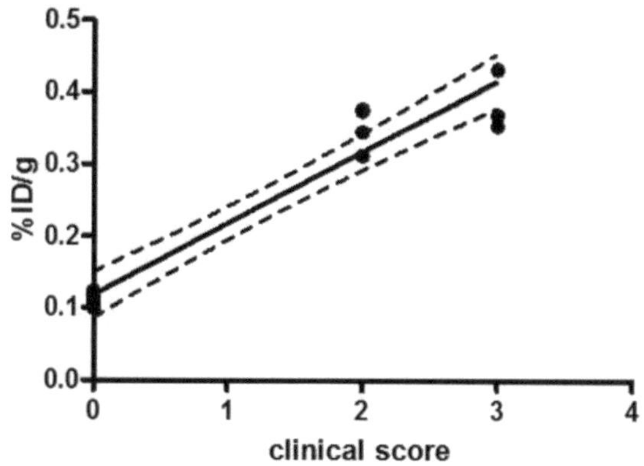

Fig. 2. Correlation between uptake of [^{123}I]-CLINDE and disease severity in the lumbar spinal cord. Linear regression analysis, r squared value (0.97) and error bars being 95% confidence intervals for fit.

Organ	Control		tEAE			
	Naive n = 4	PDD n = 4	Score 0 Preclinical n = 4	Score 2 n = 4	Score 3 n = 4	Score 0 recovery n = 4
Liver	0.27±0.05	0.27±0.03	0.32±0.03	0.32±0.03	0.33±0.02	0.25±0.03
Spleen	3.01±0.45	3.12±0.34	3.49±0.10	3.14±0.32	3.24±0.37	3.33±0.49
Kidney	2.07±0.36	1.68±0.07	2.24±0.11	2.03±0.13	2.33±0.37	1.97±0.25
Lungs	2.27±0.41	2.01±0.18	2.41±0.32	2.12±0.16	2.27±0.38	2.05±0.24
Heart	3.24±0.49	2.85±0.29	3.81±0.14	2.87±0.08	3.16±0.31	3.32±0.03
Blood	0.04±0.01	0.04±0.01	0.06±0.00	0.04±0.01	0.05±0.01	0.04±0.00
Pancreas	0.61±0.12	0.51±0.04	0.64±0.06	0.58±0.07	0.62±0.05	0.55±0.08
Thymus	0.84±0.09	0.81±0.34	1.17±0.26	0.85±0.24	1.33±0.34	1.12±0.23
Adrenals	7.69±1.27	6.38±0.62	7.86±0.37	7.00±0.93	7.22±1.33	8.49±2.10

Table 1. Biodistribution of [^{123}I]-CLINDE in the visceral organs of Lewis rats given MBP-specific T lymphoblasts and controls. Results are expressed as %ID/g ± SD of wet tissue

The specificity of [^{123}I]-CLINDE binding was demonstrated in a competition study with the PK11195 (Sigma-RBI) injected in rats with tEAE clinical score 2. The *in vivo* specificity of [^{123}I]-CLINDE was tested in two groups of four rats with tEAE clinical score 2 by injecting only [^{123}I]-CLINDE or by injecting PK11195 (5 mg/kg) prior to [^{123}I]-CLINDE. PK11195 was dissolved in saline with 5% dimethyl sulfoxide and injected 5 minutes prior to the injection of 0.70 MBq of the radiotracer. Animals were sacrificed 3 h p.i. and the tissues were analysed as described above.

Administration of PK11195 reduced the uptake of [123I]-CLINDE in the CNS (Figure 3) and the visceral organs with TSPO expression (data not shown) by 65-85% confirming the specificity of the ligand.

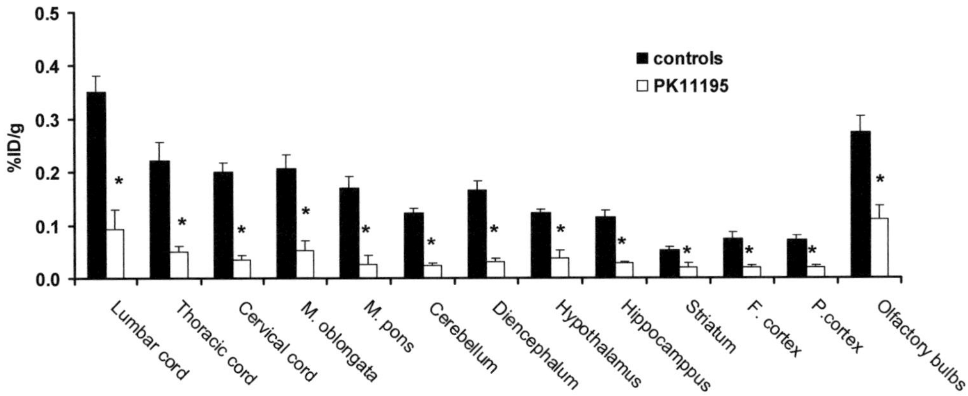

Fig. 3. Specificity of [123I]-CLINDE binding in CNS. Data are represented as mean ± SD. * $p<0.05$ - Mann Whitney test

The feasibility of using TSPO as a marker to assess the degree of microglia activation in the living rat was tested using [123I]-CLINDE on a dedicated small animal SPECT imaging system. Longitudinal imaging studies were performed using a small animal dual-head SPECT/CT camera (Flex X-SPECT, Gamma Medica Ideas Inc.) equipped with 1 mm-aperture pinhole collimators (McElroy et al., 2002). Four rats given MBP-specific T lymphoblasts and 2 control naive animals were used. The experimental rats were imaged four times (baseline, day 4, 6 and 11 post tEAE induction), while controls were imaged twice on alternate days.

Rats were injected via the tail vein with [123I]-CLINDE (20-30 MBq in 0.1 ml saline) and anaesthetized with isoflurane (2.5%). Their heads were carefully positioned at the centre of the field-of-view, and scanned between 40 and 104 minutes after the radiotracer injection. Sixty-four 1 min-projections were collected over 360 degrees by each head, at a radius-of-rotation of 45 mm. CT scan of the rat head was then performed in order to provide anatomical landmarks for the analysis. The SPECT projections of the two heads were combined and reconstructed with an iterative cone-beam algorithm (16 subsets, 4 iterations). The reconstructed data were then scaled using calibration factors to allow the measured activity to be expressed as %ID/mL. SPECT and CT volumes were automatically fused for each of the eight scans. All CT volumes of the same rat were then manually co-registered to the first one (reference), using an image visualisation and processing software (http://brainvisa.info). A rat brain atlas was finally coregistered onto the reference volume and statistics derived for each Region of Interest.

The average values of [123I]-CLINDE uptake in the brain measured before the induction of tEAE (expressed as a percentage of injected dose per mL) (0.64 ± 0.1%ID/mL) is equivalent to the average uptake of all controls imaged during the study period (0.71 ± 0.3%ID/mL). The peak increase in activity is reached 6 days post immunization and the uptake depends on the clinical score of the rat. The uptake of [123I]-CLINDE, as assessed by SPECT imaging during the course of the study, is exemplified in the hypothalamus and midbrain (Fig 4A). The spatial distribution of the activity in the brain of one typical rat over time is shown in figure 4B.

Using this small animal SPECT system we could not find statistically significant differences between intra-cerebral structures in the different groups of animals. This was probably due to the fluctuation of uptake values among animals from the same experimental group showing different tEAE scoring (Figure 4A). Moreover based on the example in Figure 4, the peak uptake in different structures of the brain differs from rat to rat depending on the tEAE score, disease severity, and therefore averaging the values for the experimental group did not reach statistical significance in the data. However the brain structures that showed an increase in uptake by SPECT imaging were the same structures where increased uptake was observed in the biodistribution studies.

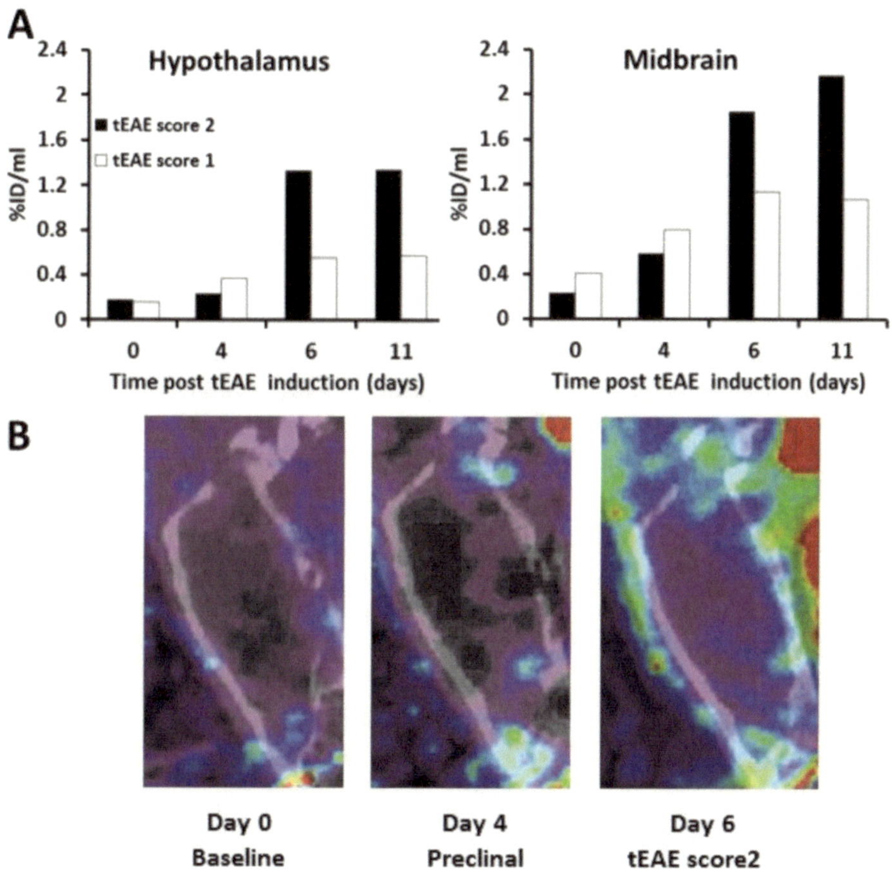

Fig. 4. *In vivo* SPECT imaging of TSPO in brain of rats given MBP-specific T lymphoblasts. **(A)** Representative individual dynamics of [^{123}I]-CLINDE uptake in hypothalamus and midbrain (expressed as a percentage of injected dose per mL). **(B)** Representative SPECT images of [^{123}I]-CLINDE uptake in a rat brain. The SPECT images shown were coregistered with CT at 0 (baseline), 4 and 6 days after induction of tEAE.

2.2 Autoradiography and immunochemistry correlation with the upregulation of TSPO in the tEAE model

In vitro and *ex vivo* autoradiography with [^{125}I]-CLINDE was performed to delineate the CNS areas with TSPO uptake.

For *in vitro* autoradiography, animals with tEAE clinical scores: 2 (*n* = 3) and 0 (after recovery, *n* = 5) and 7 control naive rats were used. Brains and spinal cords were frozen in isopentane at -80°C. Coronal sections (20 µm thick) were cut, mounted onto polysine coated slides and stored at -20°C. TSPO receptors were labeled with [^{125}I]-CLINDE (3.7 GBq/µmol) at 3 nM concentration in Tris-HCl, pH 7.4 at 4°C for 60 min. Nonspecific binding was defined by incubating adjacent tissue sections with 10 µM of PK11195. The incubation was terminated by rinsing sections twice for 2 min in cold incubation buffer. Sections were then dipped briefly in cold distilled water and dried rapidly under a stream of cold air. The sections were affixed together with calibration standards (Amersham) to radiographic films (Amersham Hyperfilm-βmax) for 4-6 h and developed using GBX developer and fixer (Kodax). The autoradiograms were analysed using a Microcomputer Imaging Device (MCID, Imaging research, Ontario, Canada).

For *ex vivo* autoradiography rats with tEAE clinical scores: 3 (*n* = 3) and 0 (after recovery, *n* = 3) and control naive rats (*n* = 3) were given [^{125}I]-CLINDE (1.85 MBq in 100 µl of saline) i.v. and were sacrificed 3 h after the injection of the radiotracer. Brains and spinal cords were frozen in isopentane at -80°C. Coronal sections (20 µm thick) were cut, thawed, dried and affixed to radiographic films (Amersham Hyperfilm-βmax) for 10-15 days and the autoradiograms analyzed as above.

In vitro autoradiography with [^{125}I]-CLINDE of the spinal cord from rats with tEAE compared to controls showed statistically significant increased binding of TSPO in areas of the grey (2 times) and white matter (1.2 - 3 times) (Figure 5). This study extends earlier observations using [^{3}H](*R*)-PK11195 (Banati et al., 2000) in which increased binding in the spinal cords from rats with tEAE, at peak of clinical disease was documented.

The *in vitro* and *ex vivo* autoradiography revealed an increase in the TSPO expression at three brain levels (Figure 6), i.e. nuclear diagonal band/rostral migratory stream; substantia nigra cerebral peduncle and mammillary nucleus; ventral cochlear nucleus. Histologically, a lesion was observed in one of the five brains at level -3 mm to bregma, in the region of substantia nigra.

As the autoradiography showed a clear increase in TSPO labelling at specific regions in the forebrain, midbrain and hindbrain as well as in the spinal cord (Figures 5 and 6), immunohistochemistry was also performed in order to clarify the presence of gliosis in the TSPO positive areas.

For this purpose, brains from 2 naive rats, 2 rats given PPD-lymphoblasts (day 4), 4 rats given MBP-lymphoblasts (day 4), 3 rats tEAE score 2 and 3 rats after recovery from the episode, were fixed in formalin, placed in David Kopf Instruments (Tujunga, CA) brain blocker and cut at six levels at a distance of 2 mm. Serial 4 µm coronal paraffin sections were taken at each of the six levels and assessed according to the rat brain atlas of Paxinos and Watson (Paxinos & Watson, 2007).

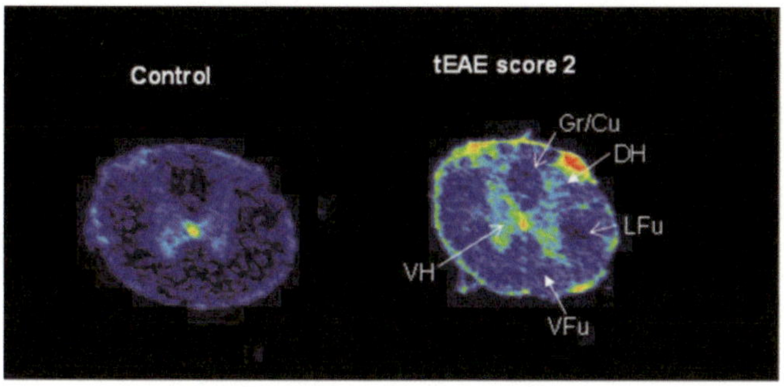

Grey matter	Control	tEAE Score 2	tEAE recovery Score 0
Dorsal horn (DH)	155 ± 17	365 ± 36***	236 72‡
Ventral horn (VH)	225 ± 17	225 ± 30	270 26
White matter			
Lateral funiculus (LFu)	123 ± 10	266 ± 23***	132 ± 31‡‡‡
Ventral funiculus (VFu)	128 ± 14	369 ± 64***	118 ± 18‡‡‡
Gracile/ Cuneate Fasciculus (Gr/Cu)	117 ± 18	228 ± 7**	113 ± 17‡‡

Fig. 5. Different cervical spinal cord TSPO expression in control rats and in rats with tEAE, using *in vitro* [^{125}I]-CLINDE autoradiography. Results are expressed as fmol/mg T.E ± SD, *n* = 3-7, one way ANOVA with Tukeys Post Hoc test (Control vs Score 2, **p<0.01, *** p<0.001; Score 2 vs Recovery, ‡ p<0.05, ‡‡ p<0.01, ‡‡‡ p<0.001)

Indeed, at the level of the midbrain, in three of the four rats studied, astrogliosis was found around the lateral ventricles on day 4 after tEAE induction, i.e. just before and at the time of onset of the clinical EAE signs (Figure 7). However, with labelling for GFAP and ED-1 no changes were observed in any of the other five brain levels studied as well as in none of the six levels in three rats with tEAE score 2 and in three rats after recovery from score 2 to score 0.

The autoradiography and immunohistochemistry results confirm the immunofluorescence observations of Meeson et al (Meeson et al., 1994) indicating very little brain inflammation that is localised in the forebrain around the third ventricle. It is tempting to speculate that the change in the medial habenular nucleus and the dentate gyrus gliosis reflects a stress reaction (Sugama et al., 2002), as this was observed around the onset of tEAE and did not persist during its development.

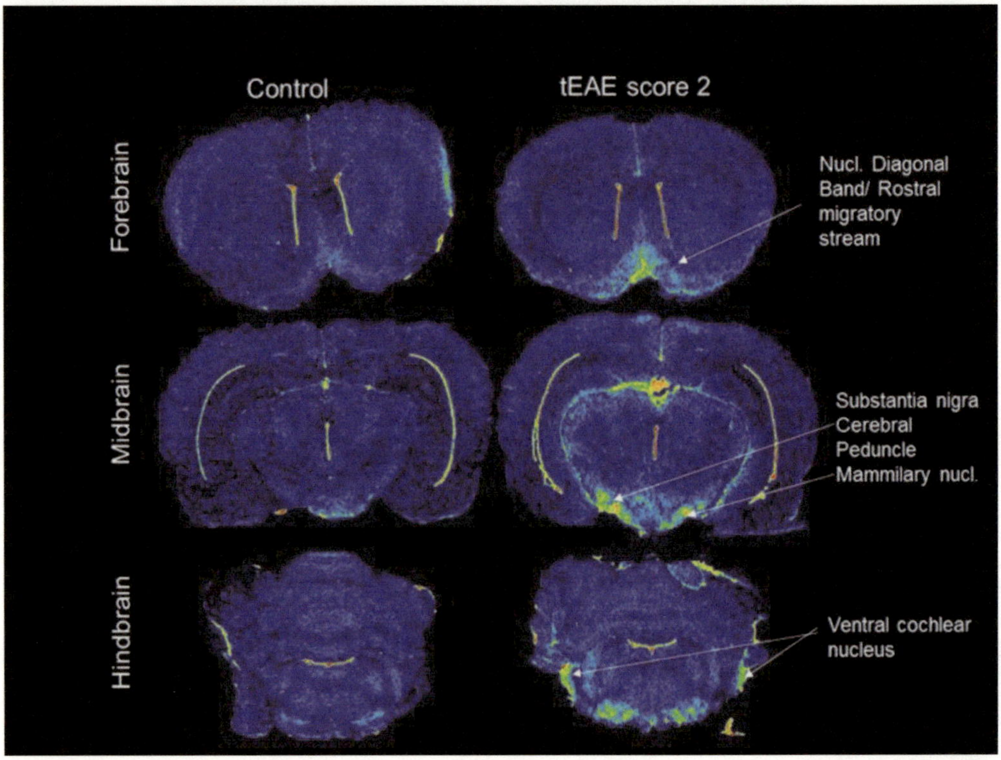

Fig. 6. Brain *in vitro* autoradiography after administration of [^{125}I]-CLINDE in Lewis rats given MBP-specific T lymphoblasts and controls

Longitudinal sections of the spinal cords were examined at the lumbar, thoracic and cervical level. After antigen retrieval (20 min boiling in citrate buffer, pH 6.0) and blocking of endogenous peroxidase and non-specific antibody binding, the macrophage/activated microglial cells were labelled with mouse anti-rat ED-1 biotin-conjugated monoclonal antibody (Serotec); TSPO - with goat polyclonal anti-TSPO IgG (Santa Cruz Biotechnology); the astrocytes - with rabbit anti-GFAP antibody (Sapphire Bioscience) and the chemokine CXCL11 - with rabbit anti-rat I-TAC IgG (kind gift from Prof. Shaun McCall, Chemokine Biology, Department of Molecular Biosciences, Adelaide University). The staining was with InnoGenex IHC Kit (San Ramon, CA) with aminoethyl carbazole (AEC) as peroxidase substrate. The sections were counterstained with Mayer's haematoxylin.

Although the histopathologic analysis and magnetic resonance (MR) microscopy had excellent correlation regarding the extent of white matter lesions on the rat EAE model (Steinbrecher et al., 2005), this imaging technology is not suitable for detection of inflammatory infiltrates in the grey matter. In our experiments significant [^{123}I]-CLINDE uptake was registered in the grey matter of the spinal cord of Lewis rats with tEAE score 2 and 3 (Figure 5). This increased uptake correlated with the positive staining for the chemokine CXCL11 that is produced mainly by astrocytes as well as with the presence of numerous ED-1 positive cells (macrophages/glial cells) that were visualised in the spinal cord meninges and parenchyma during the tEAE episode (Figure 8). Importantly, some of these activated glial cells were positive for TSPO that supports the autoradiography quantitative data. Two days after spontaneous recovery from tEAE episode, few parenchymal cells were positive for ED-1 and there was no staining for TSPO and the chemokine CXCL11 (I-TAC).

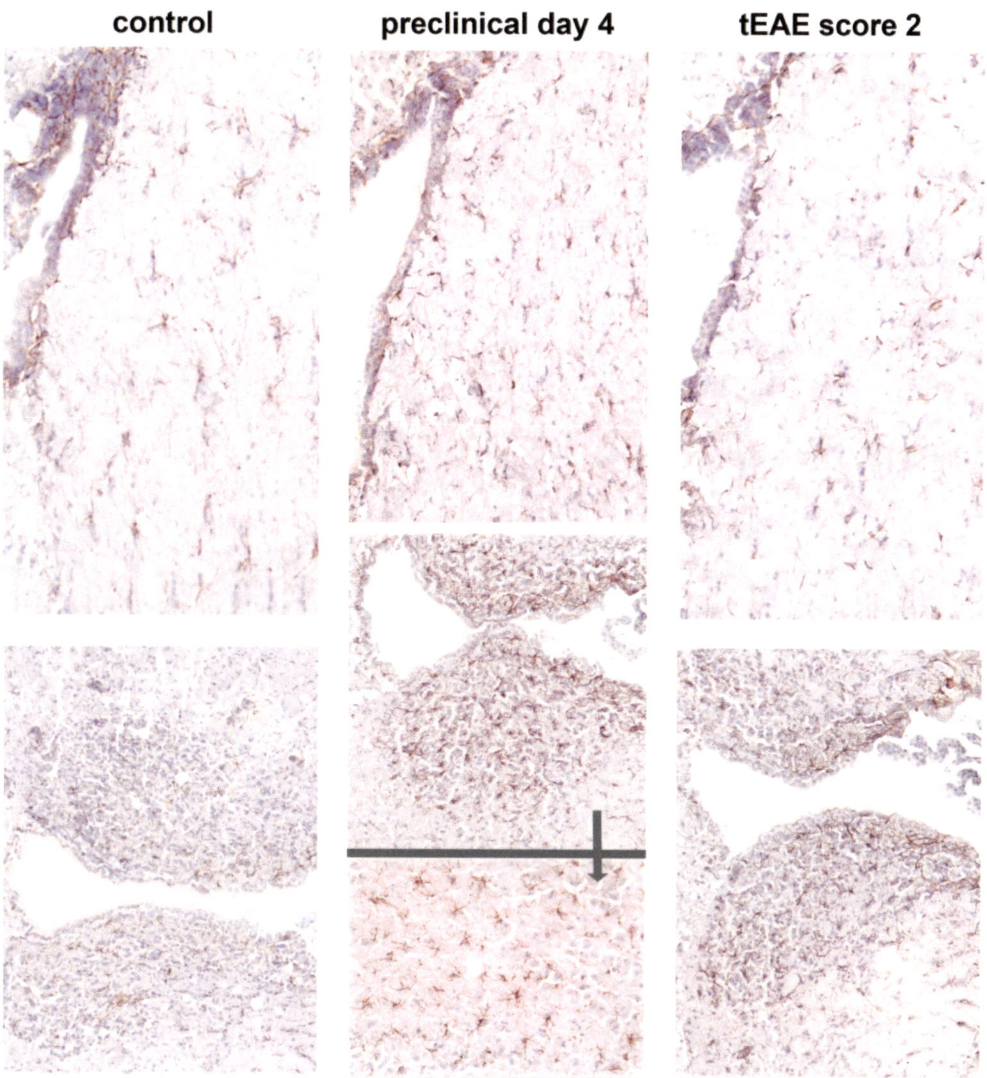

Fig. 7. Glial activation in Lewis rat tEAE (immunohistochemical evaluation). Midbrains labelled for GFAP. At disease onset, astroglyosis was observed around the 3rd ventricle in four of five animals.

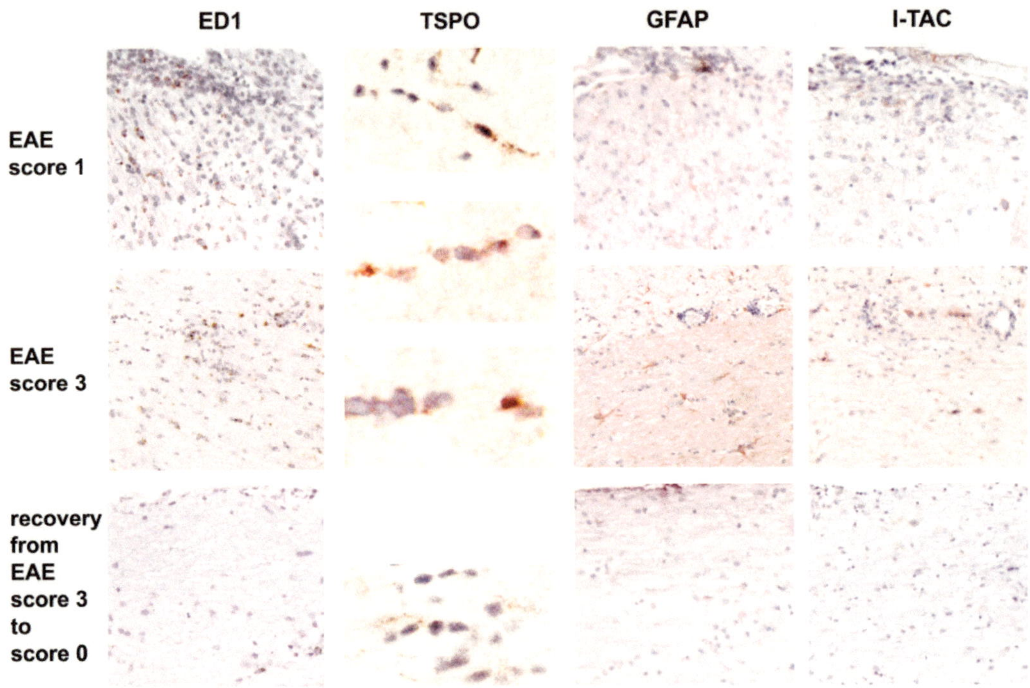

Fig. 8. Longitudinal sections of lumbar spinal cord labelled for macrophages/activated microglia (ED-1), TSPO, astrocytes and I-TAC at different clinical stages of tEAE.

One of the first *in vivo* MRI studies on tEAE in rats (Morrissey et al., 1996) showed that MRI changes were observed well before the onset of major cellular infiltration and before the onset of clinical signs that made possible to assess quantitatively the breach of the blood brain barrier (BBB) and to distinguish *in vivo* between two components of the early phase of the lesion - inflammatory infiltrates and vasogenic oedema. Increased binding of several TSPO ligands has been reported after brain injury, including focal (Myers et al., 1991a; Myers et al., 1991b) and global (Stephenson et al., 1995) cerebral ischemia in the rat where heterogeneous level of expression of TSPO in different cells is seen in the core of infarction as early as 4 days after ischemia (Rojas et al., 2007).

In this model of neuroantigen-specific neuroinflammation the significant changes were also registered from day 4 onward with SPECT imaging. A study on the same model (Banati et al., 2000) also found normal autoradiography up to day 4 with [3H]-PK11195. One should point out that we used two appropriate controls (not only naive rats). Interestingly, the time frame is similar for PET scan studies in stroke patients - the increase in binding was as early as 3 days after the onset of stroke (Price et al., 2006). Thus, any earlier in vivo as well as in vitro changes are most likely attributed to stress reaction.

The imaging studies, using the SPECT tracer [123I]-CLINDE in Lewis rat models of transferred EAE (this study) and active EAE (Mattner et al., 2005) confirm the MRI studies by other groups that the imaging changes parallel the monocyte/microglia/astrocyte activation rather than the lymphocyte infiltration (Morrissey et al., 1996; Rausch et al., 2003). These results confirm the reliability of the translocator protein (Papadopoulos &

Lecanu, 2009) in early diagnostics of antigen-specific neuroinflammation and eventually - of MS.

N.B. Discussion with mouse tEAE models was not included as they result in inflammation which progresses to demyelination (Mokhtarian et al., 1984) while the MBP-specific T lymphoblasts induced tEAE in the Lewis rat is monophasic with full recovery (Holda et al., 1980; Paterson et al., 1981).

3. Conclusion

Transferred EAE was induced in Lewis rats with MBP-specific T lymphoblasts and the uptake of the translocator protein (TSPO) tracer [^{123}I]-CLINDE was studied by biodistribution, *in vitro* and *ex vivo* autoradiography, immunohistochemistry and SPECT imaging. On a background of the typical TSPO ligand biodistribution in the visceral organs, a statistically significant 2-4 fold increase was measured in brains and spinal cords of animals with EAE clinical score of 3, compared to controls (naive or given disease-irrelevant PPD-specific T-lymphoblasts). Importantly, using [^{123}I]-CLINDE as a radiotracer we were able to register significant inflammation also in the grey matter. The CNS regional [^{123}I]-CLINDE uptake correlated with the immunohistochemical localisation of activated glial cells. The results demonstrate the ability of this highly specific TSPO ligand to measure changes in TSPO density according to area of involvement and the severity of disease suggesting it is a useful SPECT tracer for studies on experimentally induced inflammation and in multiple sclerosis.

4. Acknowledgment

The authors sincerely thank Dr David Willenborg, a Visiting Fellow at the ANU Medical School, for his encouragement and advice. M. Staykova is a Visiting Fellow at the John Curtin School of Medical Research, ANU, Canberra, Australia.

5. References

Arlicot, N.; Katsifis, A.; Garreau, L.; Mattner, F.; Vergote, J.; Duval, S.; Kousignian, I.; Bodard, S.; Guilloteau, D. & Chalon, S. (2008). Evaluation of CLINDE as Potent Translocator Protein (18 kDa) SPECT Radiotracer Reflecting the Degree of Neuroinflammation in a Rat Model of Microglial Activation. *European Journal of Nuclear Medicine and Molecular Imaging,* Vol.35, No.12, (December 2008), pp. 2203-2211, ISSN 1619-7089

Arlicot, N.; Petit, E.; Katsifis, A.; Toutain, J.; Divoux, D.; Bodard, S.; Roussel, S.; Guilloteau, D.; Bernaudin, M. & Chalon, S. (2010). Detection and Quantification of Remote Microglial Activation in Rodent Models of Focal Ischaemia Using the TSPO Radioligand CLINDE. *European Journal of Nuclear Medicine and Molecular Imaging,* Vol.37, No.12, (December 2010) pp. 2371-2380, ISSN 1619-708

Banati, R.B.; Newcombe, J.; Gunn, R.N.; Cagnin, A.; Turkheimer, F.; Heppner, F.; Price, G.; Wegner, F.; Giovannoni, G.; Miller, D.H.; Perkin, G.D.; Smith, T.; Hewson, A.K.; Bydder, G.; Kreutzberg, G.W.; Jones, T.; Cuzner, M.L. & Myers, R. (2000). The Peripheral Benzodiazepine Binding Site in the Brain in Multiple Sclerosis -

Quantitative In Vivo Imaging of Microglia as a Measure of Disease Activity. *Brain* Vol.123, (November 2000), pp. 2321-2337 ISSN 0006-8950

Boutin, H.; Chauveau, F.; Thominiaux, C.; Kuhnast, B.; Gregoire, M.C.; Jan, S.; Trebossen, R.; Dolle, F.; Tavitian, B.; Mattner, F. & Katsifis, A. (2007a). In Vivo Imaging of Brain Lesions With [C-11]ClINME, a New PET Radioligand of Peripheral Benzodiazepine Receptors. *Glia,* Vol.55, No.14, (November 2007), pp. 1459-1468, ISSN 0894-1491

Boutin, H.; Chauveau, F.; Thominiaux, C.; Gregoire, M.C.; James, M.L.; Trebossen, R.; Hantraye, P.; Dolle, F.; Tavitian, B. & Kassiou, M. (2007b). C-11-DPA-713: a Novel Peripheral Benzodiazepine Receptor PET Ligand for in Vivo Imaging of Neuroinflammation. *Journal of Nuclear Medicine,* Vol.48, No.4, (April 2007), pp. 573-581, ISSN 0161-5505

Chauveau, F.; Boutin, H.; Van Camp, N.; Dolle, F. & Tavitian, B. (2008). Nuclear Imaging of Neuroinflammation: a Comprehensive Review of [C-11]PK11195 Challengers. *European Journal of Nuclear Medicine and Molecular Imaging,* Vol.35, No.12, (December 2008), pp. 2304-2319, ISSN 1619-7070

Chen, M.K.; Baidoo, K.; Verina, T. & Guilarte, T.R. (2004). Peripheral Benzodiazepine Receptor Imaging in CNS Demyelination: Functional Implications of Anatomical and Cellular Localization. *Brain,* Vol.127, No.6, (July 2004), pp. 1379-1392, ISSN 0006-8950

Chen, M.K. & Guilarte, T.R. (2008). Translocator Protein 18 Kda (TSPO): Molecular Sensor of Brain Injury and Repair. *Pharmacology & Therapeutics,* Vol.118, No.1 (April 2008), pp. 1-17, ISSN 0163-7258

De Mestre, A.M.; Staykova, M.A.; Hornby, J.R.; Willenborg, D.O. & Hulett, M.D. (2007). Expression of the Heparan Sulfate-Degrading Enzyme Heparanase is Induced in Infiltrating CD4+ T Cells in Experimental Autoimmune Encephalomyelitis and Regulated at the Level of Transcription by Early Growth Response Gene 1. *Journal of Leukocyte Biology,* Vol.82, No.5, (November 2007) pp. 1289-1300, ISSN 0741-5400

Doorduin, J.; De Vries, E.F.J.; Dierckx, R.A. & Klein, H.C. (2008). PET Imaging of the Peripheral Benzodiazepine Receptor: Monitoring Disease Progression and Therapy Response in Neurodegenerative Disorders. *Current Pharmaceutical Design,* Vol.14, No.31, (November 2008), pp. 3297-3315, ISSN 1381-6128

Fookes, C.J.; Pham, T.Q.; Mattner, F.; Greguric, I.; Loc'h, C.; Liu, X.; Berghofer, P.; Shepherd, R.; Gregoire, M.C. & Katsifis, A. (2008). Synthesis and Biological Evaluation of Substituted [18F]Imidazo[1,2-a]pyridines and [18F]Pyrazolo[1,5-a]pyrimidines for the Study of the Peripheral Benzodiazepine Receptor Using Positron Emission Tomography. *Journal of Medicinal Chemistry,* Vol.51, No.13, (July 2008), pp. 3700-3712, ISSN 1520-4804

Gulyas, B.; Makkai, B; Kasa, P.; Gulya, K.; Bakota, L.; Varszegi, S.; Beliczai, Z.; Andersson, J.; Csiba, L.; Thiele, A.; Dyrks, T.; Suhara, T.; Suzuki, K.; Higuchi, M. & Halldin, C. (2009). A Comparative Autoradiography Study in Post Mortem Whole Hemisphere Human Brain Slices Taken from Alzheimer Patients and Age-Matched Controls Using Two Radiolabelled DAA1106 Analogues With High Affinity to the

Peripheral Benzodiazepine Receptor (PBR) System. *Neurochemistry International*, Vol.54, No.1, (January 2009), pp. 28-36, ISSN 0197-0186

Hauser, S.L.; Bhan, A.K.; Gilles, F.; Kemp, M.; Kerr, C. & Weiner, H.L. (1986). Immunohistochemical Analysis of the Cellular Infiltrate in Multiple Sclerosis Lesions. *Annals of Neurology*, Vol.19, No.6, (June 1986), pp. 578-587, ISSN 0364-5134

Holda, J.H.; Welch, A.M. & Swanborg, R.H. (1980). Autoimmune Effector Cells. I. Transfer of Experimental Encephalomyelitis with Lymphoid Cells Cultured with Antigen. *European Journal of Immunology*, Vol.10, No.8, (August 1980), pp. 657-659, ISSN 0014-2980

James, M.L.; Fulton, R.R.; Vercoullie, J.; Henderson, D.J.; Garreau, L.; Chalon, S.; Dolle, F.; Selleri, S.; Guilloteau, D. & Kassiou, M. (2008). DPA-714, a New Translocator Protein-Specific Ligand: Synthesis, Radiofluorination, and Pharmacologic Characterization. *Journal of Nuclear Medicine*, Vol.49, No.5, (May 2008), pp. 814-822, ISSN 0161-5505

Katsifis, A.; Mattner, F.; Dikic, B. & Papazian, V. (2000). Synthesis of Substituted [123I]Imidazo[1,2-a]pyridines as Potential Probes for the Study of the Peripheral Benzodiazepine Receptors Using SPECT. *Radiochimica Acta*, Vol. 88, No. 3-4, pp. 229-232, ISSN 0033-8230

Mattner, F.; Katsifis, A.; Staykova, M.; Ballantyne, P. & Willenborg, D.O. (2005). Evaluation of a Radiolabelled Peripheral Benzodiazepine Receptor Ligand in the Central Nervous System Inflammation of Experimental Autoimmune Encephalomyelitis: a Possible Probe for Imaging Multiple Sclerosis. *European Journal of Nuclear Medicine and Molecular Imaging*, Vol.32, No.5 (May 2005), pp. 557-563, ISSN 1619-7070

Mattner, F.; Mardon, K. & Katsifis, A. (2008). Pharmacological Evaluation of [I-123]-ClINDE: a Radioiodinated Imidazopyridine-3-Acetamide for the Study of Peripheral Benzodiazepine Binding Sites (PBBS). *European Journal of Nuclear Medicine and Molecular Imaging*, Vol.35, No.4, (April 2008), pp. 779-789, ISSN 1619-7070

Mattner, F.; Bandin, D.; Staykova, M.; Berghofer, P.; Gregoire, M.; Ballantyne, P.; Quinlivan, M.; Fordham, S.; Pham, T.; Willenborg, D. & Katsifis, A. (2011). Evaluation of [^{123}I]-CLINDE as a Potent SPECT Radiotracer to Assess the Degree of Astroglia Activation in Cuprizone-Induced Neuroinflammation. *European Journal of Nuclear Medicine and Molecular Imaging*, Vol. 38, No. 8 (August 2011), pp. 1516-1528, ISSN 1619-7070

McElroy, D.P.; MacDonald, L.R.; Beekman, F.J.; Yuchuan, W.; Patt, B.E.; Iwanczyk, J.S.; Tsui, B.M.W. & Hoffman, E.J. (2002). Performance Evaluation of A-SPECT: a High Resolution Desktop Pinhole SPECT System for Imaging Small Animals. *IEEE Transactions on Nuclear Science*, Vol.49, No.5 (May 2002), pp. 2139-2147, ISSN 0018-9499

McEnery, M.W.; Snowman, A.M.; Trifiletti, R.R. & Snyder, S.H. (1992). Isolation of the Mitochondrial Benzodiazepine Receptor: Association with the Voltage-Dependent

Anion Channel and the Adenine Nucleotide Carrier. *Proccedings of the National Academy of Sciences USA,* Vol.89, No.8 (May 1992), pp. 3170-3174, ISSN 0027-8424

McFarland, H.F. & Martin, R. (2007). Multiple sclerosis: a complicated picture of autoimmunity. *Nature Immunology,* Vol.8, No.9, (September 2007), pp. 913-919, ISSN 1529-2908

Meeson, A.P.; Piddlesden, S.; Morgan, B.P. & Reynolds, R. (1994). The Distribution of Inflammatory Demyelinated Lesions in the Central Nervous System of Rats with Antibody-Augmented Demyelinating Experimental Allergic Encephalomyelitis. *Experimental Neurology,* Vol.129, No.2 (October 1994), pp. 299-310, ISSN 0014-4886

Mokhtarian, F.; McFarlin, D.E. & Raine, C.S. (1984). Adoptive Transfer of Myelin Basic Protein-Sensitized T Cells Produces Chronic Relapsing Demyelinating Disease in Mice. *Nature,* Vol.309, No.5966, (May 1984), pp. 356-358, ISSN 0028-0836

Morrissey, S.P.; Stodal, H.; Zettl, U.; Simonis, C.; Jung, S.; Kiefer, R.; Lassmann, H.; Hartung, H.P.; Haase, A. & Toyka, K.V. (1996). In vivo MRI and its Histological Correlates in Acute Adoptive Transfer Experimental Allergic Encephalomyelitis. Quantification of Inflammation and Oedema. *Brain,* Vol.119, No.Pt 1, (February 1996), pp. 239-248, 0006-8950

Myers, R.; Manjil, L.G.; Cullen, B.M.; Price, G.W.; Frackowiak, R.S. & Cremer, J.E. (1991a). Macrophage and Astrocyte Populations in Relation to [3H]PK 11195 Binding in Rat Cerebral Cortex Following a Local Ischaemic Lesion. *Journal of Cerebral Blood Flow Metabolism,* Vol.11, No.2, (March 1991), pp. 314-322, 0271-678X

Myers, R.; Manjil, L.G.; Frackowiak, R.S. & Cremer, J.E. (1991b). [3H]PK 11195 and the Localisation of Secondary Thalamic Lesions Following Focal Ischaemia in Rat Motor Cortex. *Neuroscience Letters,* Vol.133, No.1, (November 1991), pp. 20-24 ISSN 0304-3940

Papadopoulos, V.; Amri, H.; Boujrad, N.; Cascio, C.; Culty, M.; Garnier, M.; Hardwick, M.; Li, H.; Vidic, B.; Brown, A.S.; Reversa, J.L.; Bernassau, J.M. & Drieu, K. (1997). Peripheral Benzodiazepine Receptor in Cholesterol Transport and Steroidogenesis. *Steroids,* Vol.62, No.1, (January 1997), pp. 21-28, ISSN 0039-128X

Papadopoulos, V.; Baraldi, M.; Guilarte, T.R.; Knudsen, T.B.; Lacapere, J.J.; Lindemann, P.; Norenberg, M.D.; Nutt, D.; Weizman, A.; Zhang, M.R. & Gavish, M. (2006). Translocator Protein (18 kDa): New Nomenclature for the Peripheral-Type Benzodiazepine Receptor Based on its Structure and Molecular Function. *Trends in Pharmacological Sciences,* Vol.27, No.8 (August 2006), pp. 402-409, ISSN 0165-6147

Papadopoulos, V. & Lecanu, L. (2009). Translocator Protein (18 kDa) TSPO: an Emerging Therapeutic Target in Neurotrauma. *Experimental Neurology,* Vol.219, No.1 (September 2009), pp. 53-57, ISSN 0014-4886

Paterson, P.Y.; Day, E.D. & Whitacre, C.C. (1981). Neuroimmunologic Diseases: Effector Cell Responses and Immunoregulatory Mechanisms. *Immunology Reviews,* Vol.55, No.1, (January 1981), pp. 89-120, ISSN0105-2896

Paxinos, G. & Watson, C. (2007). *The rat brain in stereotaxic coordinates.* Elsevier Academic Press, 6th Edition, ISBN 978-0-12-547612-6, San Diego, USA

Price, C.J.; Wang, D.; Menon, D.K.; Guadagno, J.V.; Cleij, M.; Fryer, T.; Aigbirhio, F.; Baron, J.C. & Warburton, E.A. (2006). Intrinsic Activated Microglia Map to the Peri-Infarct Zone in the Subacute Phase of Ischemic Stroke. *Stroke*, Vol.37, No.7, (July 2010), pp. 1749-1753, ISSN 1524-4628

Rausch, M.; Hiestand, P.; Baumann, D.; Cannet, C. & Rudin, M. (2003). MRI-Based Monitoring of Inflammation and Tissue Damage in Acute and Chronic Relapsing EAE. *Magnetic Resonance in Medicine,* Vol.50, No.2, (February 2003), pp. 309-314, ISSN 07403194

Rojas, S.; Martin, A.; Arranz, M.J.; Pareto, D.; Purroy, J.; Verdaguer, E.; Llop, J.; Gomez, V.; Gispert, J.D.; Millan, O.; Chamorro, A. & Planas, A.M. (2007). Imaging Brain Inflammation with [C-11]Pk11195 by PET and Induction of the Peripheral-Type Benzodiazepine Receptor After Transient Focal Ischemia in Rats. *Journal of Cerebral Blood Flow and Metabolism,* Vol.27, No.12, (December 2007), pp. 1975-1986, ISSN 0271-678X

Song, P.J.; Barc, C.; Arlicot, N.; Guilloteau, D.; Bernard, S.; Sarradin, P.; Chalon, S.; Garreau, L.; Kung, H.F.; Lantier, F. & Vergote, J. (2010). Evaluation of Prion Deposits and Microglial Activation in Scrapie-Infected Mice Using Molecular Imaging Probes. *Molecular Imaging and Biology,* Vol.12, No.6, (December 2010), pp. 576-582, ISSN 1860-2002

Steinbrecher, A.; Weber, T.; Neubergeru, T.; Mueller, A.M.; Pedre, X.; Giegerich, G.; Bogdahn, U.; Jakob, P.; Haase, A. & Faber, C. (2005). Experimental Autoimmune Encephalomyelitis in the Rat Spinal Cord: Lesion Detection with High-Resolution MR Microscopy at 17.6 T. *AJNR American Journal of Neuroradiology,* Vol.26, No.1, (January 2005), pp. 19-25, ISSN 0195-6108

Stephenson, D.T.; Schober, D.A.; Smalstig, E.B.; Mincy, R.E.; Gehlert, D.R. & Clemens, J.A. (1995). Peripheral Benzodiazepine Receptors are Colocalized with Activated Microglia Following Transient Global Forebrain Ischemia in the Rat. *Journal of Neuroscience,* Vol.15, No.7, (July 1995), pp. 5263-5274, ISSN 0270-6474

Sugama, S.; Cho, B.P.; Baker, H.; Joh, T.H.; Lucero, J. & Conti, B. (2002). Neurons of the Superior Nucleus of the Medial Habenula and Ependymal Cells Express IL-18 in Rat CNS. *Brain Research,* Vol.958, No.1 (January 2002), pp. 1-9, ISSN 0006-8993

Takano, A.; Arakawa, R.; Ito, H.; Tateno, A.; Takahashi, H.; Matsumoto, R.; Okubo, Y. & Suhara, T. (2010). Peripheral Benzodiazepine Receptors in Patients with Chronic Schizophrenia: a PET Study with C-11 DAA1106. *International Journal of Neuropsychopharmacology,* Vol.13, No.7, (August 2010), pp. 943-950, ISSN 1461-1457

Venneti, S.; Lopresti, B.J. & Wiley, C.A. (2006). The Peripheral Benzodiazepine Receptor (Translocator Protein 18 kDa) in Microglia: From Pathology to Imaging. *Progress in Neurobiology,* Vol.80, No.6 (December 2006), pp. 308-322, ISSN 0301-0082

Venneti, S.; Wang, G.J.; Nguyen, J. & Wiley, C.A. (2008). The Positron Emission Tomography Ligand DAA1106 Binds with High Affinity to Activated Microglia in Human Neurological Disorders. *Journal of Neuropathology and Experimental Neurology,* Vol.67, No.10 (October 2008), pp. 1001-1010, ISSN 0022-3069

Vowinckel, E.; Reutens, D.; Becher, B.; Verge, G.; Evans, A.; Owens, T. & Antel, J.P. (1997). PK11195 Binding to the Peripheral Benzodiazepine Receptor as a Marker of Microglia Activation in Multiple Sclerosis and Experimental Autoimmune Encephalomyelitis. *Journal of Neuroscience Research*, Vol.50, No.2 (October 1997), pp. 345-353, ISSN 0360-4012

Willenborg, D.O.; Sjollema, P. & Danta, G. (1986). Immunoregulation of Passively Induced Allergic Encephalomyelitis. *Journal of Immunology,* Vol.136, No.5, (March 1986) pp. 1676-1679, ISSN 0022-1767

4

The Role of CCR7-Ligands in Developing Experimental Autoimmune Encephalomyelitis

Taku Kuwabara[1], Yuriko Tanaka[1], Fumio Ishikawa[1], Hideki Nakano[2] and Terutaka Kakiuchi[1,*]
[1]*Department of Immunology, Toho University School of Medicine*
[2]*Laboratory of Respiratory Biology, National Institute of Environmental Health Sciences, National Institute of Health*
[1]*Japan*
[2]*USA*

1. Introduction

Multiple sclerosis is a chronic, inflammatory, and demyelinating disease of the central nervous system characterized by the pathological infiltration of autoreactive leukocytes. Experimental autoimmune encephalomyelitis serves as a disease model for human multiple sclerosis in mouse and rat (Conlon et al., 1999). Experimental autoimmune encephalomyelitis is induced through sensitization with neuroantigens such as myelin oligodendrocyte glycoprotein that activates neuroantigen-reactive T cells in the peripheral lymphoid organs. These T cells subsequently migrate into the central nervous system and encounter endogenous neuroantigens, which reactivates them and leads to nerve demyelination. Thus, induction of encephalitogenic T cells and their migration into the central nervous system are critical for development of experimental autoimmune encephalomyelitis.

$CD4^+$ helper T cells secreting IFN-γ (Th1 cells) were long considered to be the predominant T cell subset inducing experimental autoimmune encephalomyelitis (Kuchroo et al., 2002; El-behi et al., 2010). This view was challenged by the finding that IFN-γ-deficient mice showed more severe experimental autoimmune encephalomyelitis than wild type mice (Ferber et al., 1996; Gran et al., 2002). More recently, IL-17-producing T helper cells (Th17 cells) have emerged as a critical pathogenic T cell subset causing experimental autoimmune encephalomyelitis or human multiple sclerosis (Langrish et al., 2005). Th17 cells produce the pro-inflammatory cytokines IL-17A, IL-17F, and IL-22 (Ghilardi and Ouyang, 2007), and mice lacking expression of IL-17 were resistant to the induction of experimental autoimmune encephalomyelitis (Komiyama et al., 2006). Recently, Th17 cells were demonstrated to disrupt the blood-brain barrier by the action of IL-17A (Huppert et al., 2010). Based on the many investigations on the encephalitogenic T cells, current concept is that both Th1 and Th17 cells participate in the development of EAE (El-behi et al, 2010). The

* Present address: Department of Advanced and Integrated Analysis of Infectious Diseases, Toho University School of Medicine

induction of pathogenic T cells appears dependent on the coordinated migration of several cell types, a phenomena regulated by chemokines (Elhofy et al., 2002). Indeed, many chemokines have been shown to be critical for the development of experimental autoimmune encephalomyelitis (Rebenko-Moll et al., 2006). As we will discuss later, chemokines CCL19 and CCL21 regulate induction of pathogenic T cells independent of their role in the migration of immune cells. These CCR7-ligand chemokines contribute for the generation of pathogenic Th17 cells which are more efficient for the induction of experimental autoimmune encephalomyelitis.

Entry of primed T cells into the central nervous system is governed by both integrin-dependent adhesion to blood vessels and chemokine-driven migration through the blood-brain barrier. Many chemokines have been shown to be critical for the migration of activated and propagated pathogenic T cells into the central nervous system (Rebenko-Moll et al., 2006). Among them, chemokine CCL20, a ligand for CCR6, is constitutively expressed on epitherial cells of choroid plexus in mice and humans and provides ports of lymphocytes expressing a chemokine receptor CCR6 characteristic of Th17 cells (Reboldi et al., 2009). Recently, CXCL12, a ligand for CXCR7 and CXCR4, has been shown to restrict the central nervous system entry of CXCR4-expressing leukocytes, and loss of CXCL12 from abluminal surfaces of the blood-brain barrier is critical for migration of pathogenic lymphocytes into the parenchyma of the central nervous system during inducing experimental autoimmune encephalomyelitis (Cruz-Orengo et al., 2011). CCL19 and CCL21, ligands for CCR7, also have been detected at the blood-brain barrier, and suggested their involvement in CCR7-dependent lymphocyte recruitment into the central nervous system (Alt et al., 2002).

We previously identified a spontaneous mutation in mice characterized by a defect in homing of naïve T cells to the lymph node, Peyer's patches, and splenic white pulp (paucity of lymph node T cells mice; *plt/plt* mice). These mice lack the expression of CCL19 and CCL21-ser and exhibit a migration defect in T cells and dendritic cells into the T cell zone in the secondary lymphoid organs. These mice, as well as $CCR7^{-/-}$ mice, provide a good tool for the investigation of the role of these chemokines in in vivo immune response. Using *plt/plt* mouse, we have analyzed the role CCL19 and CCL21 in the regulation of immune response (Nakano et al., 1997, 1998, 2009; Gunn et al., 1999; Vassileva et al., 1999; Nakano and Gunn, 2001; Mori et al., 2001; Yasuda et al., 2007; Kuwabara et al., 2009; Aritomi et al., 2010). Unexpectedly, in vivo $CD4^+$ T cell response is not decreased, but rather enhanced. When *plt/plt* mice were immunized with a protein antigen ovalbumin with complete Freund's adjuvant, both expansion of ovalbumin-responding $CD4^+$ T cells in the draining lymph nodes and an in vitro recall response are prolonged and do not decline for a long time as compared with those in wild type mice.

Thus, there are two opposite possibilities; *plt/plt* mice with C57BL/6 background are resistant because of the lack of the expression of CCR7-ligands at the blood-brain barrier, or quite sensitive to the induction of experimental autoimmune encephalomyelitis because of the enhanced induction of pathogenic T cells. Using *plt/plt* mice as well as CCR7-/- mice, we investigated the role of CCR7-ligands in developing experimental autoimmune encephalomyelitis. As described below, we found *plt/plt* mice with C57BL/6 background are resistant to the induction of experimental autoimmune encephalomyelitis. This resistance is due to the failure to induce pathogenic Th17 cells because of deficient IL-23 production by dendritic cells, which results from lacking expression of CCL19 and CCL21.

2. CCL19 and CCL21 are required for the development of encephalomyelitis through generation of IL-23-dependent Th17 cells

For the development of experimental autoimmune encephalomyelitis, we used C57BL/6 wild type mouse and C57BL/6-*plt/plt* mouse. They were immunized following a standard protocol for induction of experimental autoimmune encephalomyelitis, that is, subcutaneous injection with myelin oligodendrocyte glycoprotein 35-55 peptide in complete Freund's adjuvant, and subsequent intravenous injection on day 0 and day 2 with pertussis toxin (Kuwabara et al., 2009).

2.1 *plt/plt* mouse is resistant to the induction of experimental autoimmune encephalomyelitis

When C57BL/6 mice were immunized under the standard immunization protocol as described above, wild type mice developed experimental autoimmune encephalomyelitis with 100% disease incidence with onset at day 14 and the peak at 4th week after immunization, whereas *plt/plt* mice failed to develop the disease during 42 days following immunization (Figure 1, upper left panel). Confirming CCR7-ligands requirement in the disease development, similarly treated CCR7$^{-/-}$ mice did not develop experimental autoimmune encephalomyelitis (Figure 1, upper right panel). That experimental autoimmune encephalomyelitis did not develop in *plt/plt* mice might be due to the failure of pathogenic T cells to migrate into the central nervous system because of the lack of CCR7-ligands expression, as suggested previously (Alt et al., 2002). To examine this possibility, 9 days after subcutaneous immunization draining lymph node cells from wild type mice were incubated for 3 days with myelin oligodendrocyte glycoprotein 35-55 peptide, and then CD4^{+} T cells were adoptively transferred intravenously into wild type and *plt/plt* mice. As shown in Figure 1, lower panel, both wild type and *plt/plt* recipients developed experimental autoimmune encephalomyelitis with 100% disease incidence with similar clinical scores and time courses. As expected, draining lymph node cells from immunized *plt/plt* mice did not develop experimental autoimmune encephalomyelitis in naïve wild type mice (Figure 1, lower panel). These results indicated that pathogenic T cells are able to infiltrate the central nervous system to induce experimental autoimmune encephalomyelitis despite the absence of CCR7-ligands but strongly suggest that pathogenic cells fail to be generated in *plt/plt* mice immunized with myelin oligodendrocyte glycoprotein 35-55 peptide.

Thus, the dependency of experimental autoimmune encephalomyelitis development on CCR7-ligands is not due to a defect in the migration of pathogenic T cells in *plt/plt* mice, since adoptive transfer of pathogenic CD4^{+} T cells prepared from draining lymph node cells of wild type mice results in the disease development in *plt/plt* and wild type recipient mice with similar time course and disease severity.

2.2 Deficient IL-17 and IFN-γ production by draining lymph node cells from mice lacking expression of CCR7-ligands

To examine whether pathogenic cells were generated in *plt/plt* mice, we compared the in vitro recall responses of draining lymph node cells from primed wild type and *plt/plt* mice. Draining lymph node cells were prepared 9 days after immunization when experimental

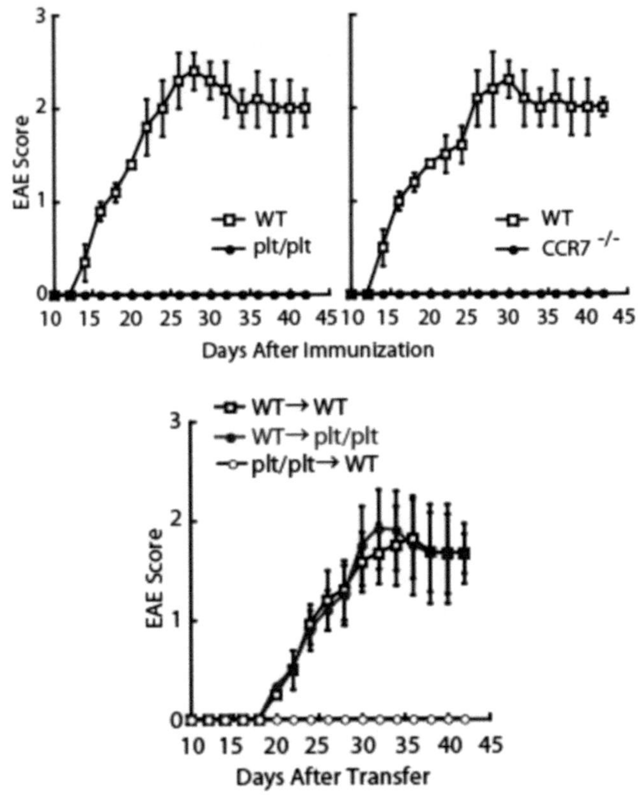

Fig. 1. Failure of *plt/plt* mice and $CCR7^{-/-}$ mice to develop experimental autoimmune encephalomyelitis. *Upper panels,* Mice were subcutaneously immunized with myelin oligodendrocyte glycoprotein 35-55 peptide in complete Freund's adjuvant at flanks and intravenously injected with pertussis toxin on days 0 and 2 (10 mice/group). Clinical symptoms were monitored for 42 days after immunization. Mean clinical score ± SD is shown. Results from wild type and *plt/plt* mice are shown in the left panel and those from wild type and $CCR7^{-/-}$ mice in the right panel. *Lower panel,* Draining lymph node cells were prepared from wild type or *plt/plt* mice 9 days after immunization and incubated with myelin oligodendrocyte glycoprotein 35-55 peptide for 3 days. Wild type $CD4^+$ T cells or *plt/plt* $CD4^+$ T cells ($1x10^7$) prepared from the treated cells were intravenously transferred into naïve and 500R X-irradiated wild type or *plt/plt* mice (10 mice/group). Results are shown as mean experimental autoimmune encephalomyelitis clinical score ± SD. WT: wild type. (Kuwabara et al., 2009)

autoimmune encephalomyelitis symptoms were not observed in wild type mice, and 14 days after immunization when the symptoms became evident. The proliferative recall responses to various doses of myelin oligodendrocyte glycoprotein 35-55 peptide were similar between draining lymph node cells from wild type and *plt/plt* mice prepared 9 days after and 14 days after immunization, suggesting T cell responses were similarly elicited in wild type and *plt/plt* mice. We also analyzed recall cytokine production to myelin oligodendrocyte glycoprotein 35-55 peptide. IL-4 and IL-10 were similarly produced by draining lymph node cells from wild type and *plt/plt* mice. Dose-dependent production of

IFN-γ or IL-17 was detected in cultures of draining lymph node cells from wild type and plt/plt mice, but production of each of these cytokines was severely diminished in *plt/plt* draining lymph node (Figure 2). These results suggest *plt/plt* T cells could be primed by immunization with myelin oligodendrocyte glycoprotein 35-55 peptide, but that the pattern of cytokine responses differed from wild type mice.

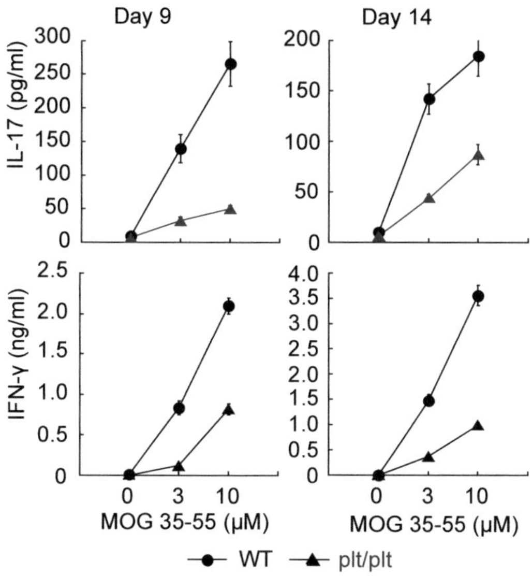

Fig. 2. *In vitro* response to myelin oligodendrocyte glycoprotein 35-55 peptide of draining lymph node cells from wild type and *plt/plt* mice. Wild type and *plt/plt* mice were immunized, as described in the legend for Figure 1. Draining lymph node cells prepared 9 or 14 days after immunization were incubated with myelin oligodendrocyte glycoprotein 35-55 peptide at indicated doses, and assessed for IFN-γ, and IL-17 in the culture supernatants by enzyme-linked immunosorbent assay using OptEIA kits (BD Biosciences). Each result is expressed as mean ± SD. (Kuwabara et al., 2009)

2.3 Requirement for CCR-7 ligands in the generation of IL-17- or IFN-γ-secreting T cells

Reduced in vitro IL-17 and IFN-γ production by draining lymph node cells from *plt/plt* mice suggested a defect in Th17 and Th1 cell generation. To examine this possibility, draining lymph node cells were prepared 9 days after immunization, incubated with myelin oligodendrocyte glycoprotein 35-55 peptide and assessed for intracellular IL-17 or IFN-γ staining. As shown in Figure 3, $CD4^{+}IL\text{-}17^{+}$ Th17 cells were found at a much lower frequency in draining lymph node cells from *plt/plt* mice than in those from wild type mice (0.4% vs. 4.2%). Addition of CCL19 or CCL21 to DLN cells from *plt/plt* mice during incubation with myelin oligodendrocyte glycoprotein 35-55 peptide restored Th17 cell generation from 0.4% to 3.0 or 4.1%, respectively (Figure 3). Also the frequency of $CD4^{+}IFN\text{-}\gamma^{+}$ Th1 cells was much lower in *plt/plt* mice than in WT mice (0.4% vs. 4.4%). Addition of CCL19 or CCL21 restored Th1 cell generation in *plt/plt* mouse draining lymph node cells from 0.4% to 3.1 or 3.2%, respectively (Figure 3). These results support the hypothesis that the defect in generating Th17 or Th1 cells in *plt/plt* mice was due to the lack of CCR7-ligand expression.

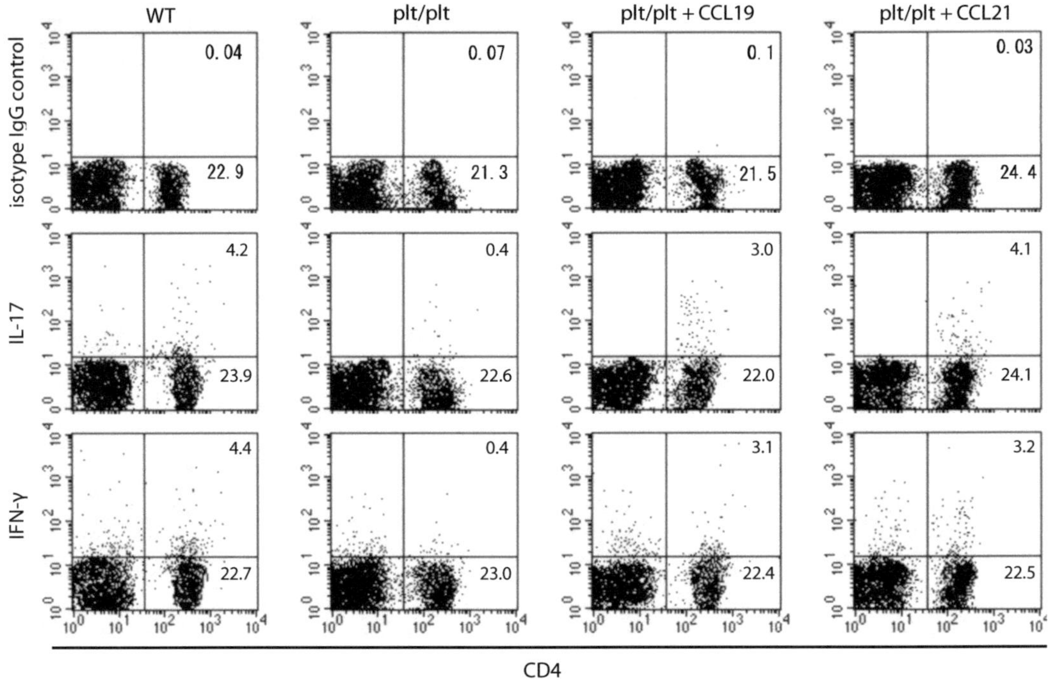

Fig. 3. Analysis of the T cell response in draining lymph nodes from wild type and *plt/plt* mice immunized for experimental autoimmune encephalomyelitis induction and generation of Th17 or Th1 cells by CCR7-ligand. Draining lymph node cells were prepared from wild type and *plt/plt* mice 9 days after immunization as described in the legend for Figure 1. Draining lymph node cells were incubated with myelin oligodendrocyte glycoprotein 35-55 peptide in the presence or absence of CCL21 or CCL19 (100ng/ml) then assessed for intracellular IL-17 or IFN-γ expression on a flow cytemeter. Numbers in right quadrants are the percentage to the total cells. (Kuwabara et al., 2009)

2.4 Decreased production of IL-12 and IL-23 by draining lymph node cells from *plt/plt* mice

For the optimal induction of IL-17-producing cells, IL-6, TGF-β and IL-23 are required (Veldhoen et al., 2006; Bettelli et al., 2006; Mangan et al., 2006). IL-12 is critical for inducing IFN-γ-producing cells (Seder and Paul, 1994) . Deficient production of IL-17 and IFN-γ suggested that these cytokines were insufficiently produced in draining lymph node cells from *plt/plt* mice. Draining lymph node cells prepared from wild type and *plt/plt* mice 4 or 9 days after immunization, similar levels of IL-6 and TGF-β production were observed following incubation with myelin oligodendrocyte glycoprotein 35-55 peptide. In contrast, as shown in Figure 4, the expression of IL-23p19 mRNA and IL-12p35 mRNA and production of IL-23 and IL-12 were much lower in cells from *plt/plt* mice than wild type mice, suggesting that the defect in production of IL-17 and IFN-γ in draining lymph node cells from *plt/plt* mice was due to insufficient provision of IL-23 and IL-12.

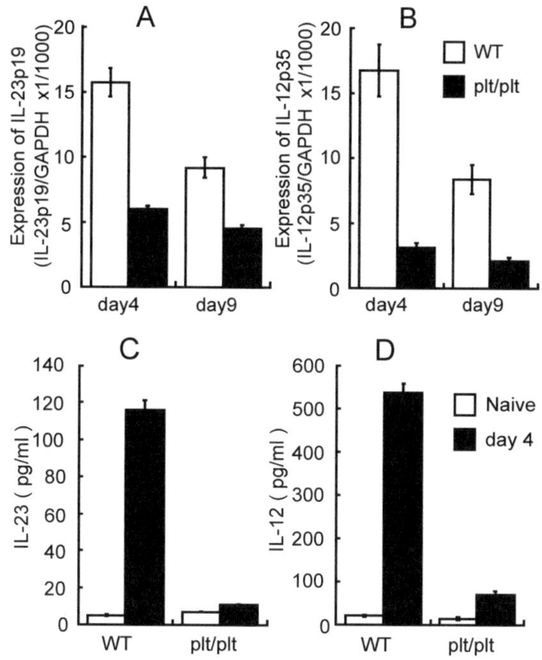

Fig. 4. Severely impaired production of IL-12 and IL-23 in draining lymph node cells from *plt/plt* mice. Draining lymph node cells were prepared 4 or 9 days after immunization as described in the legend for Figure 1. **A, B,** Expression of IL-23p19 mRNA (A) and IL12p35 mRNA (B) in $CD11c^+$ cells was estimated by quantitative RT-PCR in draining lymph node cells from wild type and *plt/plt* mice. The expression is shown as mean ± SD of the ratio to GAPDH, an internal control. These experiments were repeated 5 times with similar results. **C, D,** Draining lymph node cells from naïve mice or 4days after immunization were incubated with 10 μM myelin oligodendrocyte glycoprotein 35-55 peptide for 24 hrs. Culture supernatants were assessed for IL-23 (C) and IL-12 (D). Results of triplicate assay were presented as mean ± SD. (Kuwabara et al., 2009)

2.5 Th17 cells critically participate in the development of experimental autoimmune encephalomyelitis

Previous reports demonstrated that neuroantigen-specific Th17 or Th1 cell is responsible for experimental autoimmune encephalomyelitis induction (Langrish et al., 2005; Lees et al., 2008; Kroenke et al., 2008). To determine which defect in generating Th17 or Th1 cells was more critical in the resistance to experimental autoimmune encephalomyelitis development, draining lymph node cells from plt/plt mice were stimulated in vitro with myelin oligodendrocyte glycoprotein 35-55 peptide under the conditions for generating Th17 cells or Th1 cells, enriched for $CD4^+$ T cells, and transferred into wild type mice. As shown in Figure 5, $CD4^+$ T cells containing Th17 cells ($CD4^+IL\text{-}17^+$cells: 9.2%, $CD4^+IFN\text{-}\gamma^+$ cells: 0.1%) induced experimental autoimmune encephalomyelitis in the recipient mice with 100% disease incidence, whereas those containing Th1 cells ($CD4^+IL\text{-}17^+$cells: 0.1%, $CD4^+IFN\text{-}\gamma^+$ cells: 11.0%) did not, indicating that Th1 cells are less efficient at inducing experimental autoimmune encephalomyelitis, at least under the conditions employed. The cell preparation containing Th17 or Th1 cells was confirmed to predominantly produce IL-17 or IFN-γ, respectively. The cells similarly prepared from WT mice and enriched for Th1 cells ($CD4^+$ $IL\text{-}17^+$cells: 0.6%, $CD4^+$ IFN-γ^+ cells: 20.3%) also failed to elicit experimental autoimmune encephalomyelitis in the recipient mice, whereas those containing Th17 cells ($CD4^+$ $IL\text{-}17^+$cells: 19.2%, $CD4^+$ IFN-γ^+ cells: 0.4%) elicited experimental autoimmune encephalomyelitis. These findings strongly support our interpretation that the defect in generating Th17 cells is crucial in the resistance to experimental autoimmune encephalomyelitis development in *plt/plt* mice under the conditions employed.

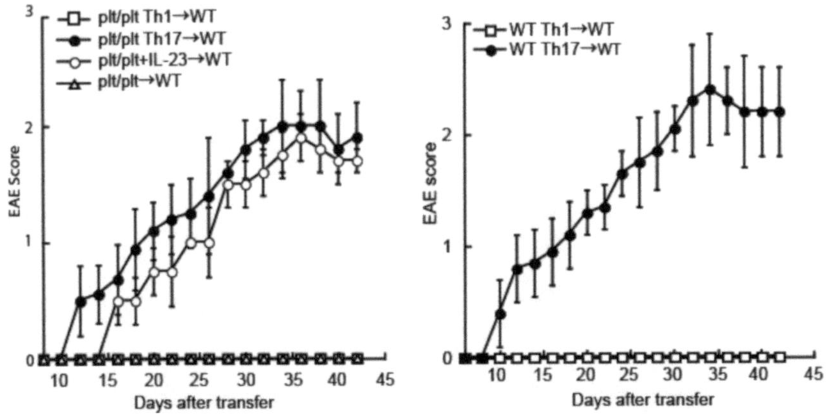

Fig. 5. Th17-enriched, rather than Th1-enriched, cell population was responsible for experimental autoimmune encephalomyelitis development in recipient mice. Draining lymph node cells from primed *plt/plt* (left panel) or wild type (right panel) mice were incubated with myelin oligodendrocyte glycoprotein 35-55 peptide for 3 days in the presence of CCL19, IL-12 and anti-IL-4 and anti-IL-23 mAbs for developing Th1 cells, in the presence of CCL19, IL-23 and anti-IL-4 and anti-IFN-γ mAbs for developing Th17 cells, or in the presence of IL-23 alone. $CD4^+$ T cells ($1x10^7$) prepared from the treated cells were intravenously transferred into naïve and 500R X-irradiated wild type mice (10 mice/group). Experimental autoimmune encephalomyelitis development is shown as a mean clinical score ± SD. (Kuwabara et al., 2009)

2.6 IL-23-dependent induction of encephalitogenic Th17 cells

Deficient IL-23 production in draining lymph node cells from *plt/plt* mice prompted us to evaluate the role of IL-23 in inducing Th17 cells. Addition of exogenous IL-23 to $CD4^+$ draining lymph node cells from immunized *plt/plt* mice stimulated with immobilized anti-CD3 and anti-CD28 mAbs increased the frequency of Th17 cells from 0.18% to 1.34%, supporting the idea that the defect in developing Th17 cells in *plt/plt* mice was due to reduced production of IL-23. To confirm that stimulation with IL-23 was able to induce pathogenic T cells in experimental autoimmune encephalomyelitis induction, draining lymph node cells from immunized *plt/plt* mice were incubated with myelin oligodendrocyte glycoprotein 35-55 peptide in the presence of IL-23, enriched for $CD4^+$ T cells, and adoptively transferred into naïve wild type mice, which resulted in the development of experimental autoimmune encephalomyelitis in the recipient mice (Figure 5, left panel). These results suggested that exogenous IL-23 was able to stimulate *plt/plt* mouse draining lymph node cells along with myelin oligodendrocyte glycoprotein 35-55 peptide to induce pathogenic Th17 cells, consistently, with the critical role of IL-23 in the induction phase of experimental autoimmune encephalomyelitis (Thankker et al., 2007). Taken all together, these findings suggest that the defect in *plt/plt* mice is likely a defect in Th17 cell generation due to deficient IL-23 production.

2.7 CCR7-ligands stimulate dendritic cells to produce IL-23

Dendritic cells are known to produce IL-23 (Oppmann et al., 2000). The reduced production of IL-23 in the incubation of *plt/plt* draining lymph node cells with myelin oligodendrocyte glycoprotein 35-55 peptide suggests the dependency of the IL-23 production on CCR7-ligands. To confirm this possibility, we prepared bone marrow-derived dendritic cells and stimulated the cells with CCR7-ligands or other chemokines. Lipopolysaccharide was used as a positive control for induction of IL-23p19 mRNA (Oppmann et al., 2000). CCL19 or CCL21 increased IL-23p19 mRNA expression, although not to the same extent as lipopolysaccharide (Figure 6-A, left and middle panels). The chemokines CCL5 and CXCL12 did not stimulate bone marrow-derived dendritic cells to produce IL-23 (Figure 6, left panel). Confirming that CCL19 and CCL21 stimulate DCs through CCR7 to express IL-23p19mRNA, bone marrow-derived dendritic cells from CCR7$^{-/-}$ mice did not respond to the chemokines (Figure 6-A, right panel).

Draining lymph node cells also express IL-23p19 mRNA in response to CCR7-ligands. Draining lymph node cells from immunized wild type, *plt/plt*, or CCR7$^{-/-}$ mice were incubated with myelin oligodendrocyte glycoprotein 35-55 peptide for 6 hours in the presence or absence of CCL19 or CCL21. Then, CD11c^{+} cells were enriched and assayed for IL-23p19 mRNA expression. As shown in Figure 6-B, left panel, CD11c^{+} cells from wild type mice expressed much higher IL-23p19 mRNA than those from naïve mice, and addition of CCL19 did not further enhance IL-23p19 mRNA expression in these cells from immunized wild type mice, probably because they had been exposed to CCL19 produced in draining lymph nodes. In CD11c^{+} cells from *plt/plt* mice, however, addition of exogenous CCL19 or CCL21 increased IL-23p19 mRNA expression (Figure 6-B, middle). As expected, cells from CCR7$^{-/-}$ mice did not respond to the addition of CCR7-ligands (Figure 6-B, right panel).

CCR7-ligands also stimulated IL-23 production by bone marrow-derived dendritic cells from wild type and *plt/plt* mice and by draining lymph node cells from *plt/plt* mice (Figure 6-C, D). Draining lymph node cells alone from immunized wild type mice produced much more IL-23 than those from naïve wild type mice, probably because endogenous CCR7-ligands induced sufficient level of IL-23 production (Figure 6- C, D). Taken together, the results shown in Figure 6 demonstrate that CCL19 or CCL21 is necessary and sufficient to induce IL-23 production from dendritic cells. Confirming that IL-23 production in response to a CCR7-ligand plays a critical role in Th17 induction, in a dose-dependent fashion anti-IL-23 mAb inhibited Th17 cell generation following incubation of draining lymph node cells from *plt/plt* mice with myelin oligodendrocyte glycoprotein 35-55 peptide in the presence of CCL19 or CCL21 (Figure 7).

Also in vivo expression of IL-23 in dendritic cells was observed in the presence of CCR7-ligands, but not in the absence of them. When mice were immunized subcutaneously with a protein antigen ovalbumin in complete Freund's adjuvant, expression of IL-23p19 mRNA

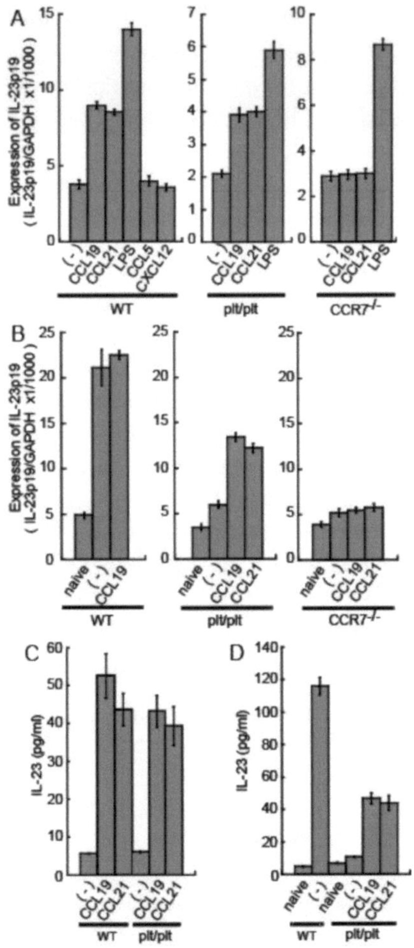

Fig. 6. CCR7-ligands stimulate dendritic cells to express IL-23p19 mRNA and to produce IL-23. **A,** Bone marrow-derived dendritic cells were prepared from wild type, *plt/plt*, and CCR7$^{-/-}$ mice, and stimulated with lipopolysaccharide or indicated chemokines at 100ng/ml for 6 hours. Cellular RNA was prepared from each cell population and IL-23p19 mRNA expression was evaluated by quantitative RT-PCR. The expression is shown as mean ± SD of the ratio to GAPDH, an internal control. **B,** Draining lymph node cells were prepared 4 days after immunization from wild type, *plt/plt*, and CCR7$^{-/-}$ mice, and incubated with 10μM myelin oligodendrocyte glycoprotein 35-55 peptide in the presence or absence of CCL19 or CCL21 for 6 hrs. CD11c^{+} cells were enriched with a positive selection kit (BD Biosciences) by MACS. CD11c^{+} cells were 89.2%, 92.2%, and 90.4% for wild type, *plt/plt*, and CCR7$^{-/-}$ mice, respectively. Cellular RNA was prepared from each cell population and assessed for IL-23p19 mRNA expression by quantitative RT-PCR. Controls were lymph node cells from naïve mice. Expression is shown as mean ± SD of the ratio to GAPDH as an internal control. **C, D,** Bone marrow-derived dendritic cells (C) or draining lymph node cells (D) from wild type and *plt/plt* mice were stimulated as described above for 24 hours. The supernatants were assessed for IL-23 using an enzyme-linked immunosorbent assay kit (BD Biosciences). Results are shown as a mean ± SD of triplicate assay. (Kuwabara et al., 2009)

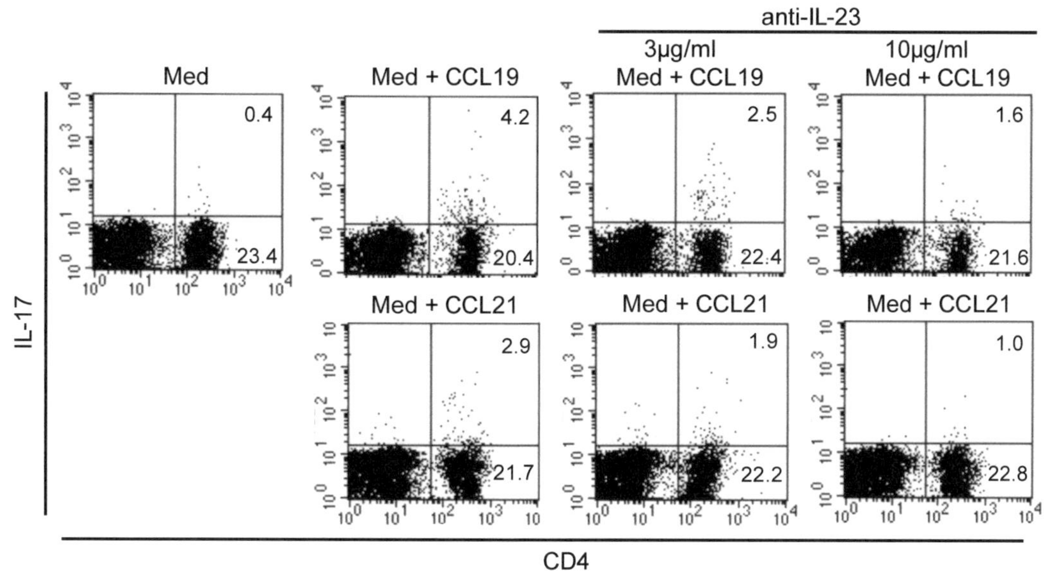

Fig. 7. Draining lymph node cells from *plt/plt* mice were incubated with myelin oligodendrocyte glycoprotein 35-55 peptide for 2 days in the presence or absence of CCL21 or CCL19 alone or with anti-IL23 mAb. The cells were analyzed for CD4 expression and intracellular IL-17. (Kuwabara et al., 2009)

and its protein was much higher in the draining lymph nodes CD11c+ dendritic cells from wild type mice than in those from *plt/plt* mice. Thus, CCR7-ligands are required for IL-23 production both in vivo and in vitro.

IL-12p35 mRNA expression and IL-12 production in bone marrow-derived dendritic cells from *plt/plt* mice were also induced by the addition of exogenous CCL19 or CCL21.

It was also possible CCR7-ligands directly stimulated CD4+ T cells to produce IL-17. However, this seemed unlikely since CD4+ T cells isolated from naïve *plt/plt* mice or *plt/plt* mice primed with myelin oligodendrocyte glycoprotein 35-55 peptide were not induced to produce IL-17 in response to immobilized anti-CD3 and anti-CD28 mAbs in the presence of exogenously added CCL19 or CCL21. We concluded CCR7-ligands stimulated dendritic cells to produce IL-23, which in turn resulted in Th17 differentiation. Consistently, IL-23 has been shown to be a critical Th17 growth, survival and pathogenesis-inducing factor (Verdhoen et al., 2006; Bettelli et al., 2006; Mangan et al., 2006; Ghoreschi et al., 2010).

2.8 CCR7-ligands promote the generation of pathogenic Th17 cells

To determine the pathogenicity of draining lymph node T cells from *plt/plt* mice that had been incubated with CCR7-ligands under experimental autoimmune encephalomyelitis inducing conditions, 9 days after immunization cells were incubated for 3 days with myelin oligodendrocyte glycoprotein 35-55 peptide in the presence of CCL19 or CCL21. CD4+ T cells were enriched from the treated cells and intravenously transferred into naïve wild type mice. As shown in Figure 8, the recipient mice developed experimental autoimmune encephalomyelitis with more than 70% disease incidence.

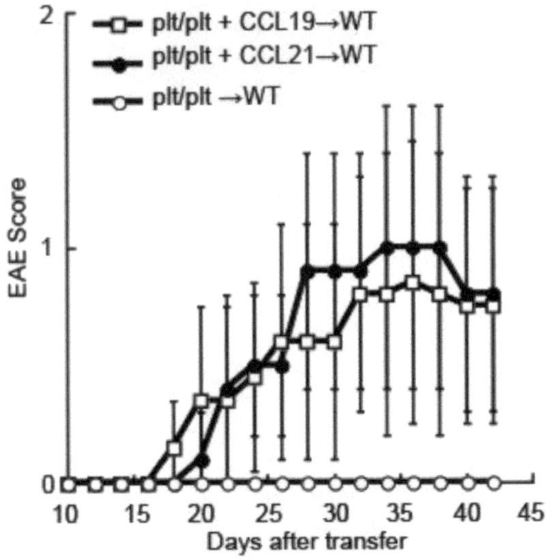

Fig. 8. Restoration of pathogenic T cells by incubation with CCR7-ligands. Draining lymph node cells from immunized *plt/plt* mice ($CD45.2^+$) were incubated at $4x10^6$ cells/ml with 10 μM myelin oligodendrocyte glycoprotein 35-55 peptide in the presence of CCR7-ligands (100 ng/ml) for 3 days. $CD4^+$ T cells ($3x10^7$) prepared from the treated cells were intravenously transferred into naïve and 500R X-irradiated wild type mice, and mice were monitored for experimental autoimmune encephalomyelitis (10 mice/group). A mean ± SD of experimental autoimmune encephalomyelitis clinical score is plotted. The experimental autoimmune encephalomyelitis incidence was 0% for recipients of cells incubated in the absence of CCR7-ligands, 70% for those in the presence of CCL19, 80% for those in the presence of CCL21. (Kuwabara et al., 2009)

2.9 CCR7 ligands up-regulate IL-23 through PI3-kinase and NF-κB pathway in dendritic cells

Finally, we explored the molecular mechanism involved in CCR7-ligand-induced IL-23 production in dendritic cells, using $CD11c^+$ spleen and bone marrow-derived dendritic cells. Although IL-23 is a heterodimeric molecule of a p40 subunit and a p19 subunit, p19 expression is the rate-limiting factor for IL-23 production (Oppmann et al., 2000). Several reports have shown that MAPK and PI3K/Akt signaling pathways triggered by CCR7 activation modulate dendritic cell function (Yanagawa & Onoe, 2002, 2003; Sanchez-Sanchez et al., 2004; Iijima et al., 2005; Riol-Blanco et al., 2005). Similar to previous studies, stimulation of dendritic cells with CCL19 or CCL21 resulted in the activation of Erk1/2, JNK, p38 MAP kinase and PI3K (Kuwabara et al., 2011). The CCR7 ligand-induced increase in IL-23 p19mRNA transcription was markedly antagonized only by a PI3K inhibitor. In contrast, the ability of dendritic cells to migrate toward CCL19 or CCL21 was not blunted by the PI3K inhibitor, indicating that signaling pathways triggered by CCR7 for IL-23 production and for migration are different (Kuwabara et al., 2011).

PI3K/Akt activation is known to induce NF-κB activation (Kane et al., 1999). Lipopolysaccharide activates NF-κB in dendritic cells to produce IL-23 (Utsugi, et al., 2006; Mise-Omata, et al.,2007; Varmody, et al., 2007; Liu, et al., 2009). We examined if NF-κB activation was also critical for CCR-7-mediated IL-23 production. When dendritic cells were stimulated with CCL19 or CCL21, translocation of NF-κB was observed from the cytoplasm into the nucleus. IκBα is an NF-κB inhibitor whose levels are inversely and closely correlated to the activation of NF-κB (Karin & Ben-Nerah). We found stimulation of dendritic cells with CCL19 or CCL21 degraded IκBα, which was prevented by inhibition of PI3K/Akt signaling. In addition, NF-κB inhibitors blunted the ability of CCR7 ligands to induce IL-23 production. Inhibition of PI3K activation abolished CCR7 ligand-mediated NF-κB DNA binding activities (Kuwabara et al., 2011). Thus, CCR7 ligands triggers NF-κB activation through PI3K/Akt signaling, which results in the production of IL-23. It was also confirmed that CCR7 ligand-stimulated dendritic cells induce Th17 cells as antigen presenting cells.

3. Conclusions

We have investigated the role of CCR7-ligands, CCL19 and CCL21, in the development of experimental autoimmune encephalomyelitis, a disease model for human multiple sclerosis in mice. For this aim we used *plt/plt* mouse lacking expression of CCL19 and CCL21-ser, which we previously identified. These mice are resistant to the induction of experimental autoimmune encephalomyelitis under the standard protocol. In these mice encephalitogenic Th17 cells are not generated. For the generation of Th17 cells IL-23 is required but dendritic cells in these mice are unable to produce IL-23. CCR7 ligands stimulate dendritic cells to produce IL-23, and dendritic cells treated with CCR7 ligands are able to generate Th17 cells as antigen-presenting cells. The molecular mechanism involved in CCR7 ligand-induced IL-23 production in dendritic cells was analyzed. CCR7 ligands trigger PI3K/Akt signaling pathway in dendritic cells through CCR7 and activate NF-κB, which results in the production of IL-23. The signaling pathway for IL-23 production is different from that for migration toward CCR7 ligands. For the development of strategies to treat experimental autoimmune encephaslomyelitis or human multiple sclerosis, we have to elucidate precise mechanisms for IL-23 production in dendritic cells through CCR7 and how dendritic cells are stimulated with CCR7 ligands in vivo to produce IL-23.

4. Acknowledgments

This work was supported in part by Project Research of Toho University School of Medicine (T. Ku., and F. I.), the Research Promotion Grants from Toho University Graduate School of Medicine (No. 05-02 to T. Ka., No.07-02 to T. Ku., No.08-02 to Y. T. and No.10-02 to T.Ku.), Grants-in-Aid for Scientific Research from the Japan Society for the Promotion of Science to T. K. (Nos. 12670621, 14021121, 17590900, and 19591013), to T. Ku. (Nos. 18790605 and 22790945), and to Y. T. (Nos. 19790695 and 21790963), and for Research on Health Sciences Focusing on Drug Innovation from the Japan Health Sciences Foundation to T.K. (KH51052), and a grant from the Japan Rheumatism Foundation to T.Ku.. We would like to thank Dr. T. Hasegawa (Ohno Chuo Hospital, Ichikawa, Japan) for his support.

5. References

Alt C, Laschinger M & Engelhardt B. (2002). Functional expression of the lymphoid Chemokines CCL19 (ELC) and CCL 21 (SLC) at the blood-brain barrier suggests their involvement in G-protein-dependent lymphocyte recruitment into the central nervous system during experimental autoimmune encephalomyelitis. *Eur J Immunol* 32:2133-44.

Aritomi K, Kuwabara T, Tanaka Y, Nakano H, Yasuda T, Ishikawa F, Kurosawa H & Kakiuchi T. (2010). Altered antibody production and helper T cell function in mice lacking chemokines CCL19 and CCL21-Ser. *Microbiol Immunol* 54:691-701.

Bettelli E, Carrier Y, Gao W, Korn T, Strom TB, Oukka M, Weiner HL & Kuchroo VK. (2006). Reciprocal developmental pathways for the generation of pathogenic effector TH17 and regulatory T cells. *Nature* 441:235-238.

Carmody R J, Ruan Q, Liou HC & Chen YH. (2007). Essential roles of c-Rel in TLR-induced IL-23 p19 gene expression in dendritic cells. *J Immunol* 178: 186-191.

Conlon P, Oksenberg JR, Zhang J & Steinman L. (1999). The immunobiology of multiple sclerosis: an autoimmune disease of the central nervous system. *Neurobiol Dis* 6:149-166.

Cruz-Orengo L, Holman DW, Dorsey D, Zhou L, Zhang P, Wright M, McCandless EE, Patel JR, Luker GD, Littman DR, Russell JH & Klein RS. (2011). CXCR7 influences leukocyte entry into the CNS parenchyma by controlling abluminal CXCL12 abundance during autoimmunity. *J Exp Med* 208:327-39.

El-behi M, Rostami A & Ciric B. (2010). Current views on the roles of Th1 and Th17 cells in experimental autoimmune encephalomyelitis. *J Neuroimmune Pharmacol* 5:189-197.

Ferber IA, Brocke S, Taylor-Edwards C, Ridgway W, Dinisco C, Steinman L, Dalton D & Fathman CG. (1996). Mice with a disrupted IFN-γ gene are susceptible to the induction of experimental autoimmune encephalomyelis (EAE). *J Immunol* 156:5-7.

Elhofy A, Kennedy KJ, Fife BT, Karpus WJ. (2002). Regulation of experimental autoimmune encephalomyelitis by chemokines and chemokine receptors. *Immunol Res* 25:167-75.

Ghilardi N & Ouyang W. (2007). Targeting the development and effector functions of Th17 cells. *Semin Immunol* 19:383-393.

Ghoreschi K, Laurence A, Yang XP, Tato CM, McGeachy MJ, Konkel JE, Ramos HL, Wei L, Davidson TS, Bouladoux N, Grainger JR, Chen Q, Kanno Y, Watford WT, Sun HW, Eberl G, Shevach EM, Belkaid Y, Cua DJ, Chen W, O'Shea JJ. (2010). Generation of pathogenic T(H)17 cells in the absence of TGF-β signalling. *Nature* 467:967-71.

Gran B, Zhang GX, Yu s, Li J, Chen XH, Ventura ES, Kamoun M & Rostami A. (2002). IL-12p35-deficient mice are susceptible to experimental autoimmune encephalomyelitis: evidence for redundancy in the IL-12 system in the induction of central nervous system autoimmune demyelination. *J Immunol* 169-7110.

Gunn MD, Kyuwa S, Tam C, Kakiuchi T, Matsuzawa A, Williams LT, Nakano H. (1999). Mice lacking expression of secondary lymphoid organ chemokine have defects in lymphocyte homing and dendritic cell localization. *J Exp Med* 189:451-460.

Huppert J, Closhen D, Croxford A, White R, Kulig P, Pietrowski E, Bechmann I, Becher B, Luhmann HJ, Waisman A & Kuhlmann CR. (2010). Cellular mechanisms of IL-17-induced blood-brain barrier disruption. *FASEB J* 24:1023-34.

Iijima N, Yanagawa Y, Clingan JM & Onoe K. (2005). CCR7-mediated c-Jun N-terminal kinase activation regulates cell migration in mature dendritic cells. *Int Immunol* 17: 1201-1212.

Kane LP, Shapiro VS, Stokoe D & Weiss A. (1999) Induction of NF-κB by the Akt/PKB kinase. *Curr Biol* 9: 601-604.

Karin M & Ben-Neriah Y. (2000). Phosphorylation meets ubiquitination: the control of NF-κB activity. *Annu Rev Immunol* 18:621-663.

Komiyama Y, Nakae S, Matsuki T, Nambu A, Ishigame H, Kakuta S, Sudo K, & Iwakura Y. (2006) IL-17 plays an important role in the development of experimental autoimmune encephalomyelitis. *J Immunol* 177:566-73.

Kuchroo VK, Anderson AC, Walddner H, Munder M, Bettelli E & Nicholson LB. (2002). T cell response in experimental autoimmune encephalomyelitis (EAE): role of self and cross-reactive antigens in shaping, tuning, and regulating the autopathogenic T cell repertoire. *Annu Rev Immunol* 20:101-123.

Kuwabara T, Ishikawa F, Yasuda T, Aritomi K, Nakano H, Tanaka Y, Okada Y, Lipp M, Kakiuchi T. (2009). CCR7 ligands are required for development of experimental autoimmune encephalomyelitis through generating IL-23-dependent Th17 cells. *J Immunol* 183:2513-21.

Kuwabara T, Tanaka Y, Ishikawa F, Kondo M, Sekiya H & Kakiuchi T. (2011). CCR7 ligands up-regulate IL-23 through PI3-kinase and NF-κB pathway in dendritic cells. *in submitted.*

Langrish CL, Chen Y, Blumenschein WM, Mattson J, Basham B, Sedgwick, JD, McClanahan T, Kastelein RA & Cua DJ. (2005). IL-23 drives a pathogenic T cell population that induces autoimmune inflammation. *J Exp Med* 201:233-240.

Liu W, Ouyang X, Yang J, Liu J, Li Q, Gu Y, Fukata M, Lin T, He JC, Abreu M, Unkeless JC, Mayer L & Xiong H. (2009). AP-1 activated by toll-like receptors regulates expression of IL-23 p19. *J Biol Chem* 284: 24006-24016.

Mangan PR, Harrington LE, O'Quinn DB, Helms WS, Bullard DC, Elson CO, Hatton RD, Wahl SM, Schoeb TR & Weaver CT. (2006). Transforming growth factor-beta induces development of the T(H)17 lineage. Nature 441:231-234.

Mise-Omata S, Kuroda E, Niikura J, Yamashita U, Obata Y & Doi TS. (2007). A proximal kappaB site in the IL-23 p19 promoter is responsible for RelA- and c-Rel-dependent transcription. *J Immunol* 179: 6596-6603.

Mori S, Nakano H, Aritomi K, Wang CR, Gunn MD & Kakiuchi T. (2001). Mice lacking expression of the chemokines CCL21-ser and CCL19 (plt mice) demonstrate delayed but enhanced T cell immune responses. *J Exp Med* 193:207-218.

Nakano H, Tamura T, Yoshimoto T, Yagita H, Miyasaka M, Butcher EC, Nariuchi H, Kakiuchi T & Matsuzawa A. (1997). Genetic defect in T lymphocyte-specific homing into peripheral lymph nodes. *Eur J Immunol* 27:215-221.

Nakano H, Mori S, Yonekawa H, Nariuchi H, Matsuzawa A & Kakiuchi T. (1998). A novel mutant gene involved in T-lymphocyte-specific homing into peripheral lymphoid organs on mouse chromosome 4. *Blood* 91:2886-2895.

Nakano H & Gunn MD. (2001). Gene duplications at the chemokine locus on mouse chromosome 4: multiple strain-specific haplotypes and the deletion of secondary lymphoid-organ chemokine and EBI-1 ligand chemokine genes in the plt mutation. *J Immunol* 166(1):361-369.

Nakano H, Lin KL, Yanagita M, Charbonneau C, Cook DN, Kakiuchi T & Gunn MD. (2009). Blood-derived inflammatory dendritic cells in lymph nodes stimulate acute T helper type 1 immune responses. *Nat Immunol* 10:394-402.

Oppmann B, Lesley R, Blom B, Timans JC, Xu Y, Hunte B, Vega F, Yu N, Wang J, Singh K, Zonin F, Vaisberg E, Churakova T, Liu M, Gorman D, Wagner J, Zurawski S, Liu Y,

Abrams JS, Moore KW, Rennick D, de Waal-Malefyt R, Hannum C, Bazan JF & Kastelein RA. (2000). Novel p19 protein engages IL-12p40 to form a cytokine, IL-23, with biological activities similar as well as distinct from IL-12. *Immunity* 13:715-725.

Rebenko-Moll NM, Liu L, Cardona A & Ransohoff RM. (2006). Chemokines, mononuclear cells and the nervous system: heaven (or hell) is in the details. *Curr Opin Immunol* 18:683-689.

Reboldi A, Coisne C, Baumjohann D, Benvenuto F, Bottinelli D, Lira S, Uccelli A, Lanzavecchia A, Engelhardt B & Sallusto F. (2009). C-C chemokine receptor 6-regulated entry of TH-17 cells into the CNS through the choroid plexus is required for the initiation of EAE. *Nat Immunol* 10:514-523.

Riol-Blanco L, Sanchez-Sanchez N, Torres A, Tejedor A, Narumiya S, Corbi AL, Sanchez-Mateos P & Rodriguez-Fernandez JL. (2005) The chemokine receptor CCR7 activates in dendritic cells two signaling modules that independently regulate chemotaxis and migratory speed. *J Immunol* 174: 4070-4080.

Sanchez-Sanchez N, Riol-BlancoL, de la Rosa G, Puig-Kroger A, Garcia-Bordas J, Martin D, Longo N, Cuadrado A, Cabanas C, Corbi A L, Sanchez-Mateos P & Rodriguez-Fernandez JL. (2004). Chemokine receptor CCR7 induces intracellular signaling that inhibits apoptosis of mature dendritic cells. *Blood* 104: 619-625.

Seder RA & Paul WE. (1994). Acquisition of lymphokine-producing phenotype by $CD4^+$ T cells. *Annu Rev Immunol* 12:635-73.

Thakker P, Leach MW, Kuang W, Benoit SE, Leonard JP & Marusic S. (2007). IL-23 is critical in the induction but not in the effector phase of experimental autoimmune encephalomyelitis. *J Immunol* 178:2589-2598.

Utsugi M, Dobashi K, Ishizuka T, Kawata T, Hisada T, Shimizu Y, Ono A & Mori M. (2006). Rac1 negatively regulates lipopolysaccharide-induced IL-23 p19 expression in human macrophages and dendritic cells and NF-κB p65 trans activation plays a novel role. *J Immunol* 177: 4550-4557.

Vassileva G, Soto H, Zlotnik A, Nakano H, Kakiuchi T, Hedrick JA & Lira SA. (1999). The reduced expression of 6Ckine in the plt mouse results from the deletion of one of two 6Ckine genes. *J Exp Med* 190:1183-1188.

Veldhoen M, Hocking RJ, Atkins CJ, Locksley RM & Stockinger B. (2006). TGF-β in the context of an inflammatory cytokine milieu supports de novo differentiation of IL-17-producing T cells. *Immunity* 24:179-89.

Yanagawa Y & Onoe K. (2002). CCL19 induces rapid dendritic extension of murine dendritic cells. *Blood* 100: 1948-1956.

Yanagawa Y & Onoe K. (2003). CCR7 ligands induce rapid endocytosis in mature dendritic cells with concomitant up-regulation of Cdc42 and Rac activities. *Blood* 101: 4923-4929.

Yasuda T, Kuwabara T, Nakano H, Aritomi K, Onodera T, Lipp M, Takahama Y & Kakiuchi T. (2007). Chemokines CCL19 and CCL21 promote activation-induced cell death of antigen-responding T cells. *Blood* 109:449-456.

Part 2

Experimental Therapeutic Approaches

5

Therapeutic Effects of the Sphingosine 1-Phosphate Receptor Modulator, Fingolimod (FTY720), on Experimental Autoimmune Encephalomyelitis

Kenji Chiba, Hirotoshi Kataoka, Noriyasu Seki and Kunio Sugahara
Mitsubishi Tanabe Pharma Corporation
Japan

1. Introduction

2-Amino-2-[2-(4-octylphenyl)ethyl]propane-1,3-diol hydrochloride (FTY720, fingolimod hydrochloride) is an orally active sphingosine 1-phosphate (S1P) receptor modulator with a structure closely related to sphingosine (Adachi et al., 1995; Chiba et al., 1996, 1997) (Fig. 1). FTY720 sequesters circulating mature lymphocytes into the secondary lymphoid organs (SLO) and thymus by long-term down-regulation of S1P receptor type 1 (S1P1) on lymphocytes, and shows potent immunomodulating effects (Brinkmann et al., 2000, 2002a, 2002b, 2004; Chiba et al., 1998, 1999, Chiba, 2005; Matloubian et al., 2004).

It has been previously reported that a potent immunosuppressive natural product, myriocin (ISP-I) can be isolated from a culture broth of *Isaria scinclairii*, a kind of vegetative wasp (Fujita et al., 1994a, 1994b; Sasaki et al.,1994). The chemical modification of ISP-I led to a novel synthetic compound, FTY720 that has more potent immunomodulating activity and less toxicity than myriocin (Adachi et al., 1995; Fujita et al., 1995, 1996; Kiuchi et al., 2000). FTY720 is highly effective in prolonging allograft survival in various experimental allograft models (Chiba et al., 1996, 1998, 2005; Hoshino et al., 1996; Kawaguchi et al., 1996; Masubuchi et al., 1996; Suzuki et al., 1996, 1998). A striking feature of FTY720 is the induction of a marked decrease in the number of peripheral blood lymphocytes (T cells and B cells) at doses that prolong allograft survival (Chiba et al., 1996, 1998, 1999; Yanagawa et al., 1998a, 1998b, 1999, 2000). The reduction of the number of peripheral blood lymphocytes induced by FTY720 is mainly caused by sequestration of circulating mature lymphocytes into the SLO as lymph nodes and Peyer's patches (Chiba et al., 1998, 1999, Yanagawa et al., 1998a, 1998b).

It has been reported that FTY720 is effectively phosphorylated to FTY720-phosphate (FTY720-P, (*S*)-enantiomer) (Fig. 1) by sphingosine kinases (Billich et al., 2003; Paugh et al., 2003) and that FTY720-P is a high affinity agonist at four types of S1P receptors (S1P1, S1P3, S1P4, and S1P5) (Brinkmann et al., 2002a, 2002b; Mandala et al., 2002). It is well documented that S1P1 plays an essential role in lymphocyte egress from the SLO to lymph (Brinkmann et al., 2004; Chiba, 2005; Matloubian et al., 2004). FTY720-P induces a long-lasting internalization and degradation of S1P1, reduces S1P responsiveness of lymphocytes, and inhibits lymphocyte egress from the SLO (Brinkmann et al 2002a, 2004; Chiba, 2005; Cyster, 2005; Graler et al., 2004, Maeda et al., 2007, Matloubian et al., 2004; Pham et al., 2008). Consequently, FTY720-P acts as a functional antagonist at lymphocytic S1P1 and immunomodulating effects of FTY720 are likely due to inhibition of egress of antigen-specific T cells from draining lymph nodes.

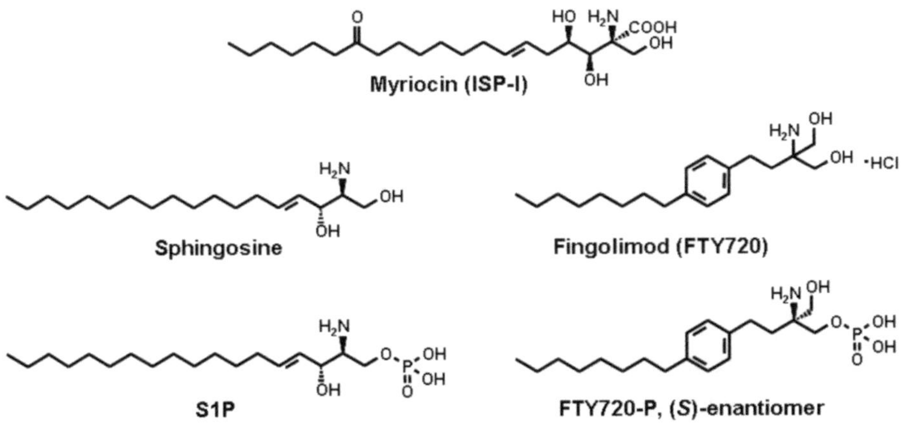

Fig. 1. The Chemical structures of myriocin (ISP-I), sphingosine, sphingosine 1-phospahete (S1P), fingolimod (FTY720) and FTY720-phosphate (FTY720-P).

Multiple sclerosis (MS) is a common and often disabling disease of the central nervous system (CNS). Early active MS lesions are characterized by the presence of infiltrated mononuclear cells around venules and small veins, followed by myelin breakdown and astrogliosis, resulting in irreversible disability (Lubin et al., 1996; Martin et al., 1992; Rammohan, 2003). The etiology of MS remains unknown, but is widely considered to involve the organ-specific autoimmune destruction of CNS myelin mediated by myelin-specific T cells (Anderson et al., 2004; Seamons et al., 2003). Immunomodulating therapy using cyclophosphamide, interferon (IFN)-β, or glatiramer acetate is widely used for the treatment of MS (Goodkin et al., 1999; Zamvil et al., 2003). Experimental autoimmune encephalomyelitis (EAE), an animal model of human MS, is a demyelinating, inflammatory disease of the CNS and is induced by the immunization of susceptible strains of rats or mice with myelin antigens combined with adjuvant (Kuchroo et al., 1993; Martin et al., 1992, 1995; Owens et al., 1995). It is highly likely that trafficking and infiltration of IFN-γ-expressing type 1 helper T cells (Th1 cells) (Bright et al., 1998; Merrill et al., 1992; Windhagen et al., 1995) and interleukin 17 (IL-17)-expressing helper T cells (Th17 cells) (Bettelli et al., 2006; Komiyama et al., 2006; Kroenke et al., 2008; Langrish et al., 2005; Steinman, 2010; Stromnes et al., 2008) into the CNS play an important role in the development and progression of EAE because myelin antigen-specific Th17 cells and Th1 cells can be found in the blood, lymphoid tissues and CNS in EAE and MS.

There are several reports on ameliorating effects of FTY720 on EAE in mice and rats (Balatoni et al., 2007; Brinkmann et al., 2002a; Chiba et al., 2011; Foster et al., 2007; Fujino et al., 2003; Kataoka et al., 2005; Webb et al., 2004). In myelin basic protein (MBP)-induced EAE in LEW rats, prophylactic administration of FTY720 at 0.1 to 1 mg/kg almost completely prevents the development of EAE symptoms and EAE-associated histological change in the spinal cords (Fujino et al., 2003; Kataoka et al. 2005). In myelin oligodendrocyte glycoprotein (MOG)-induced EAE in DA rats, prophylactic therapy of FTY720 protects against the emergence of EAE symptoms, neuropathology, and disturbances to visual and somatosensory evoked potentials (Foster et al., 2007, Balatoni et al., 2007). Consistent with rat EAE, development of EAE induced by myelin proteolipid protein (PLP) in SJL/J mice is almost completely prevented and infiltration of CD4 T cells into spinal cord is decreased by prophylactic treatment with FTY720 and FTY720-P (Webb et al., 2004; Kataoka et al., 2005). When FTY720 or FTY720-P is given after establishment of EAE in SJL/J mice, marked therapeutic effects on EAE are observed accompanying with reduction of demyelination and infiltration of CD4 T cells in the spinal cords (Chiba et al., 2011; Kataoka et al., 2005; Webb et al., 2004). Similar therapeutic effects by FTY720 are obtained in MOG-induced EAE in C57BL/6 mice (Chiba et al., 2011; Kataoka et al., 2005, Seki et al., 2010).

In this chapter, we demonstrate the prophylactic and therapeutic effects of FTY720 on EAE in rats and mice using MBP, myelin PLP or MOG as myelin antigens (Chiba et al., 2011; Kataoka et al., 2005, Seki et al., 2010). Our findings suggest that FTY720 is more efficacious in mouse EAE as compared with rm-IFN-β and that the ameliorating effects of FTY720 on EAE are likely due to reduction of the infiltration of myelin antigen-specific Th17 and Th1 cells into the CNS.

On the other hand, it has been well documented that neural cells express S1P receptors and FTY720 can distribute into the CNS beyond blood brain barrier (Brinkmann, 2009). Recently, it has been strongly suggested that FTY720-P directly acts as a functional antagonist at S1P1 on not only lymphocytes but also neural cells, particularly astrocytes because the therapeutic effects of FTY720 on EAE was lost in CNS mutants lacking S1P1 on astrocytes but not neuron (Choi et al, 2011). Since FTY720 possesses a completely new mechanism of action, FTY720 should provide a useful therapeutic approach for MS.

2. Discovery of FTY720 from a natural product, myriocin

A potent immunosuppressive natural product, myriocin was isolated from a culture broth of *Isaria sinclairii*, a kind of vegetative wasp that is an "eternal youth" nostrum in traditional Chinese herbal medicine (Fujita et al., 1994a, 1994b; Sasaki et al., 1994). Chemical modification of myriocin yielded a new compound, FTY720 (Fig. 1), which has more potent immunomodulating activity and less toxicity than myriocin (Adachi et al., 1995; Fujita et al., 1995, 1996; Kiuchi et al., 2000). FTY720 at an oral dose of 0.1 mg/kg or higher significantly prolongs allograft survival in experimental skin, cardiac and renal allotransplantation models (Chiba et al., 1996, 1997, 1998; Hoshino et al., 1996; Kawaguchi et al., 1996; Masubuchi et al., 1996; Suzuki et al., 1996, 1998). Unlike calcineurin inhibitors, FTY720 does not impair lymphocyte function including T cell activation and production of Th1 cell-associated cytokines by antigen stimulation (Chiba et al., 1996, 1997, 1998; Yanagawa et al., 1998a). Moreover, we have demonstrated that FTY720 does not inhibit serine palmitoyltransferase that is the first enzyme in sphingolipid biosynthesis and is the target enzyme of myriocin (Fujita et al., 1996).

A striking feature of FTY720 is the induction of a marked decrease in the number of peripheral blood lymphocytes at doses that display an immunomodulating activity (Chiba et al., 1996, 1997, 1998, Li et al., 2002; Yagi et al., 2000; Yanagawa et al., 1998a, 1998b). In rats, the number of lymphocytes (T cells and B cells) in peripheral blood decreases dramatically within 6 hours after oral administration of FTY720 at 0.1 to 1 mg/kg (Chiba et al., 1998, 1999). To clarify the mechanism of lymphocyte reduction by FTY720, distribution of lymphocytes in blood, lymph, and various SLO was analyzed after FTY720 administration in rats (Chiba et al., 1998). When FTY720 at an oral dose of 0.1 mg/kg or higher is given to rats or mice, the number of lymphocytes is decreased markedly in the peripheral blood and thoracic duct lymph whereas that in the SLO is increased significantly. Intravenous transfusion of fluorescein-labelled lymphocytes into rats reveals that the labelled lymphocytes are accumulated in the SLO by FTY720 administration (Chiba et al, 1998). These data suggest that FTY720 induces sequestration of circulating mature lymphocytes into SLO and decreases the number of lymphocytes in peripheral blood, and lymph. Thus, sequestration of circulating mature lymphocytes is presumed to be the main mechanism of immunomodulating activity of FTY720. As reported previously, immunohistochemical staining and flow cytomeric analysis revealed that FTY720 decreases the infiltration of T cells into the allograft at doses showing immunomodulating effects (Chiba, 2005, Yanagawa et al, 1998a, 1999, 2000). These findings strongly suggest that FTY720 exerts immunomodulating activity by decreasing T cell infiltration into inflammatory sites.

3. Mechanism of action of FTY720, S1P receptor modulator

Circulation of mature lymphocytes among the SLO, lymph and blood plays a central role in the establishment of the immune response to foreign antigens. Homing of lymphocytes from blood into the SLO beyond high endothelial venules is highly dependent on the interaction between the CC-chemokine ligand (CCL) 19, CCL21, CXC-chemokine ligand (CXCL) 13, and their receptors: CC-chemokine receptor (CCR) 7 and CXC-chemokine receptor (CXCR) 5 on lymphocytes. On the other hand, throughout the analyses of the mechanism of action of FTY720, it is clarified that S1P and its receptor, S1P1 play an essential role in lymphocyte egress from the SLO to lymph (Brinkmann et al 2002a, 2004; Chiba, 2005; Chiba et al., 2006; Cyster, 2005; Lo et al., 2005; Maeda et al., 2007, Matloubian et al., 2004; Pham et al., 2008). Reverse pharmacological approaches have been performed to elucidate that FTY720 is phosphorylated by sphingosine kinases (Billich et al., 2003; Paugh et al., 2003) and FTY720-phosphate (FTY720-P) acts as a high affinity agonist of 4 types of S1P receptors (S1P1, S1P3, S1P4, and S1P5) (Brinkmann et al, 2002a, 2002b; Mandala et al., 2002). After oral or intravenous FTY720 administration, the plasma concentration of FTY720-P was 2 to 6 times higher than FTY720 (Brinkmann et al 2002a).

S1P, a pleiotropic lysophospholipid mediator is converted primarily by the phosphorylation of sphingosine by sphingosine kinases and stimulates multiple signalling pathways resulting in calcium mobilization from intracellular stores, polymerization of actin, chemotaxis/migration, and escape from apoptosis (Hla et al., 2001, Pyne et al.; 2000). S1P is predominantly released by red blood cells or platelets and is found in significant amounts (100 to 400 nM) in serum (Kimura et al, 2001). S1P binds with nano-molar affinities to five related G-protein-coupled receptors, termed S1P1-5 (Brinkmann et al., 2002a, 2002b; Chiba, 2005; Mandala et al. 2002) (Fig. 2). It has been reported that S1P1 is essential for lymphocyte recirculation and that S1P1 regulates lymphocyte egress from the SLO (Brinkmann et al 2002, 2004; Chiba, 2005; Cyster, 2005; Maeda et al., 2007, Matloubian et al., 2004; Pham et al., 2008). In mice whose hematopoietic cells lack a single S1P receptor, S1P1, there are no T cells in the periphery because mature T cells are unable to exit SLO (Matloubian et al., 2004). Moreover, S1P1-dependent chemotactic responsiveness is suggested to be up-regulated in lymphocytes before exit from the SLO, whereas S1P1 is down-regulated during peripheral lymphocyte activation, and this is associated with retention of lymphocytes in the SLO (Lo et al, 2005).

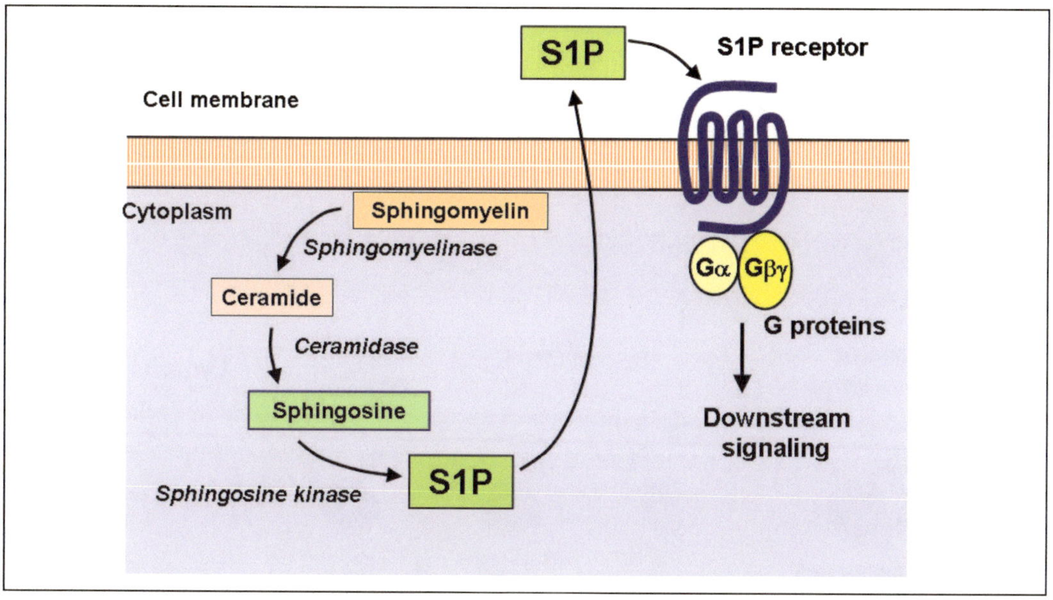

Fig. 2. Generation of S1P and interaction with S1P receptor. S1P is generated primarily by the phosphorylation of intracellular sphingosine by sphingosine kinases and binds to five related G-protein-coupled S1P receptors (S1P1-5).

FTY720 treatment down-regulates S1P1, creating a temporary pharmacological S1P1-null state in lymphocytes, providing an explanation for the mechanism of FTY720-induced lymphocyte sequestration (Graler et al., 2004; Matloubian et al., 2004). Since S1P1 surface expression on lymphocytes is highly dependent on the extracellular concentration of S1P, S1P1 on lymphocytes is down-regulated in the blood, up-regulated in SLO and down-regulated again in the lymph (Cyster, 2005; Lo et al., 2005). Thus, it is proposed that cyclical modulation of S1P1 surface expression on circulating lymphocytes by S1P contributes to establishing their transit time in SLO.

We have confirmed that only the (*S*)-enantiomer of FTY720-P can bind to four types of S1P receptors (S1P1, 3, 4, 5) at nano-molar concentrations, but not S1P2, whereas FTY720 up to 10,000 nM does not bind S1P receptors (Chiba, 2005, Kiuchi et al., 2005). FTY720-P shows agonist activity for S1P1 at nano-molar concentrations using extracellular signal regulated kinase 1/2 (ERK1/2) phosphorylation assay and subsequently induces long-term internalization of S1P1 in Chinese hamster ovary (CHO) cells stably expressing human S1P1 (Chiba, 2005; Chiba et al.; 2006, Maeda et al., 2007). The internalization of S1P1 by FTY720-P appears to be maintained longer than that by S1P and the difference between FTY720-P and S1P seems to be due to the distinct stability of FTY720-P and S1P for degradation by S1P lyase. S1P at concentrations of 10 to 100 nM induces migration of lymph node $CD4^+$ T cells in mice. By contrast, the pretreatment with FTY720-P effectively inhibits the migration of CD4+ T cells toward S1P (Chiba, 2005; Chiba et al.; 2006, Maeda et al., 2007). Based on these results, it is presumed that FTY720-P converted from FTY720 acts as a functional antagonist at S1P1 receptor by internalization and degradation of the S1P1, reduces S1P responsiveness of lymphocytes, and inhibits S1P1-dependent lymphocyte egress from SLO (Fig. 3).

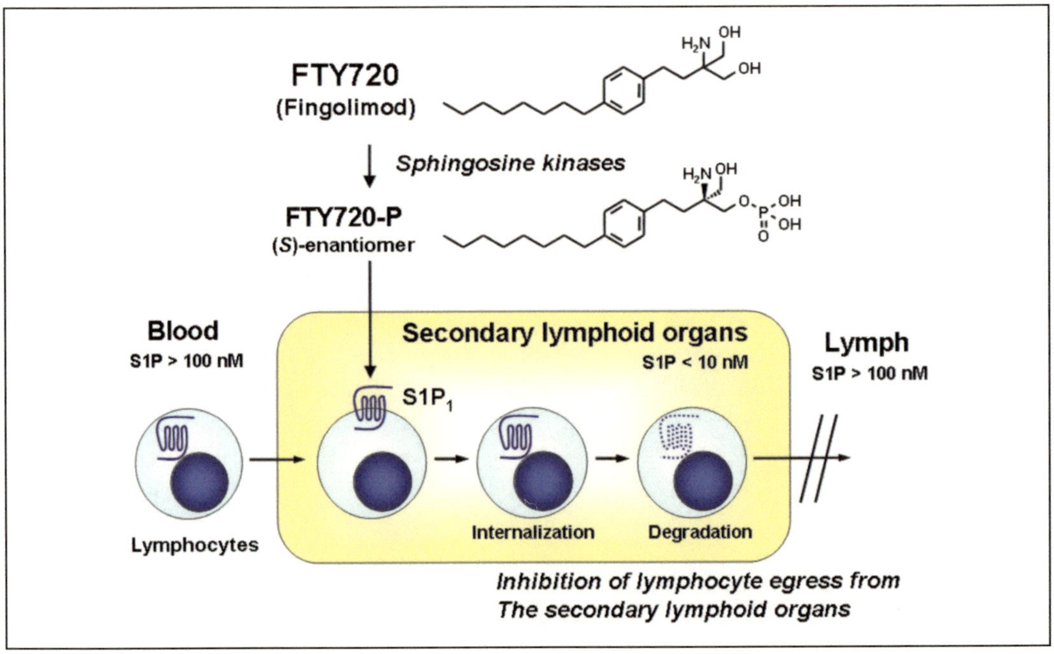

Fig. 3. FTY720-P converted from FTY720 inhibits S1P/S1P1 axis-dependent lymphocyte egress from SLO by functional antagonism (internalization and degradation) at lymphocytic S1P1.

4. Effects of FTY720 on MBP-induced EAE in LEW rats

We examined the prophylactic and therapeutic effects of FTY720 on EAE induced by immunization with MBP in LEW rats (Kataoka et al., 2005). All of the LEW rats in the vehicle-treated control group developed EAE-associated clinical symptoms 10 days after immunization with guinea pig MBP, reaching a maximal level on day 13 followed by a gradual decline (Fig. 4A). Elevation of EAE scores was significantly inhibited in groups given FTY720 prophylactically at oral doses of 0.1, 0.3, and 1 mg/kg for 20 days from the day of MBP immunization. Consistent with EAE-associated symptoms, the histological scores in FTY720-treated groups decreased significantly and dose-dependently as compared with the vehicle-treated control group.

To evaluate the therapeutic potential of FTY720 in MBP-induced EAE in LEW rats, the administration was started from the day of EAE onset (Fig. 4B). In the vehicle-treated control group, EAE had developed by day 9 after immunization and reached a maximal level on day 11 to 13. Thereafter, the mean of EAE scores remained within the 2 to 3 range until day 20, because 4 out of 8 EAE rats died with severe symptoms in the control group. Therapeutic administration of FTY720 from the day of EAE onset significantly decreased post-peak EAE-associated clinical signs. Moreover, the therapeutic administration of FTY720 resulted in a significant decrease in the histological score of EAE. The infiltration of inflammatory cells into the spinal cords of EAE rats was inhibited in the FTY720-treated group. These data indicate that FTY720 not only has a prophylactic but also a therapeutic potential in MBP-induced EAE in LEW rats.

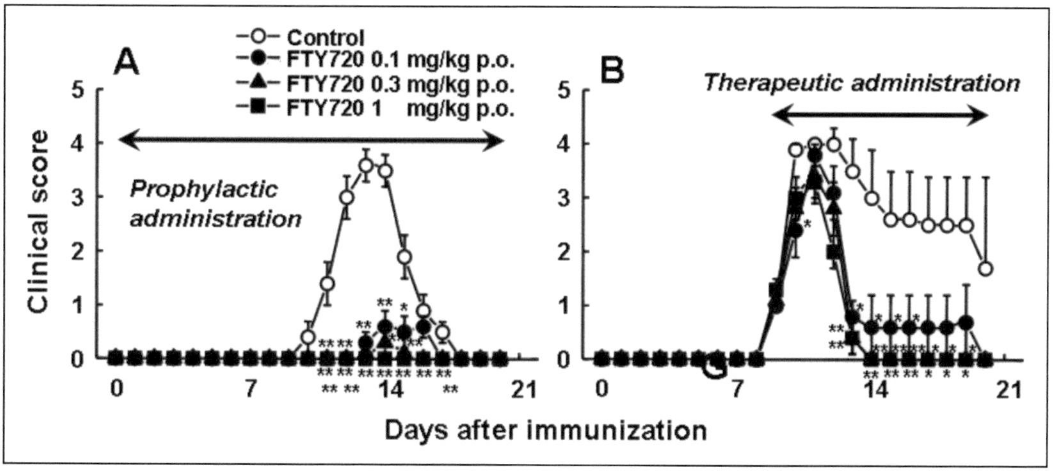

Fig. 4. Prophylactic and therapeutic effects of FTY720 on MBP-induced EAE in LEW rats. LEW rats were immunized with guinea pig MBP (100 µg/rat) and Freund's complete adjuvant. FTY720 was administered orally to MPB-immunized LEW rats every day from the day of immunization (A: prophylactic administration) or the day of onset (B: therapeutic administration) until day 20. Rats in the control group were administered vehicle only. The results are expressed as the mean ± S.E.M of 8 animals and statistical differences were calculated by Steel's test (non-parametric Dunnett's multiple comparison test, *: $p<0.05$, **: $p<0.01$).

5. Effects of FTY720 on PLP-induced EAE in SJL/J mice

5.1 Prophylactic effects of FTY720 on PLP-induced EAE in SJL/J mice

In human MS, neurological symptoms relapse over several years; however MPB-induced EAE in LEW rats was monophasic with no relapse. To clarify the therapeutic potential of FTY720 in human MS more precisely, we evaluated the effect of FTY720 and its active metabolite, FTY720-P on relapsing EAE in SJL/J mice induced by $PLP_{139-151}$ immunization (Chiba et al., 2011; Kataoka et al., 2005). SJL/J mice immunized with $PLP_{139-151}$ emulsified in Freund's complete adjuvant resulted in the development of EAE-associated clinical symptoms and a decrease in body weight 11 days after immunization. EAE scores rapidly elevated and reached a maximal level on day 15. The first phase of EAE remitted with a low EAE score on day 20 and spontaneously relapsed thereafter (Fig. 5A). The elevation of EAE-associated clinical score and the loss of body weight were prevented in groups given FTY720 prophylactically at oral doses of 0.1 and 0.3 mg/kg for 42 days from the day of immunization (Figs. 5). Consistent with MBP-induced EAE in LEW rats, there was no increase in EAE score in the group treated with FTY720 at 1 mg/kg, indicating a complete prevention of $PLP_{139-151}$-induced EAE in SJL/J mice (Kataoka et al., 2005). Almost the same prophylactic effect was observed when FTY720-P at 0.1 and 1 mg/kg was administered intraperitoneally to $PLP_{139-151}$-immunized SJL/J mice from the day of immunization. By contrast, prophylactic treatment with recombinant mouse (rm)-IFN-β at 10000 IU three times a week intraperitoneally showed no clear effect or no lymphopenia (Kataoka et al., 2005).

The infiltration of inflammatory cells was observed in the spinal cord of SJL/J mice 17 days after immunization with $PLP_{139-151}$ (Kataoka et al., 2005). Prophylactic administration of FTY720 at 1 mg/kg orally resulted in a marked reduction of the infiltration of inflammatory cells in the spinal cord of $PLP_{139-151}$-immunized SJL/J mice (Fig. 6A-D). Immunohistochemical staining analysis using anti-T cell subset mAbs revealed the infiltration of $CD4^+$ T cells rather than $CD8^+$ T cells into the spinal cord, especially the perivascular area and funiculus dorsalis in white matter under pia mater, of $PLP_{139-151}$-immunized SJL/J mice with developed EAE. Prophylactic administration of FTY720 at 1 mg/kg orally markedly decreased the infiltration of $CD4^+$ T cells into the spinal cord as compared with the vehicle-treated control EAE group (Fig. 6E-H).

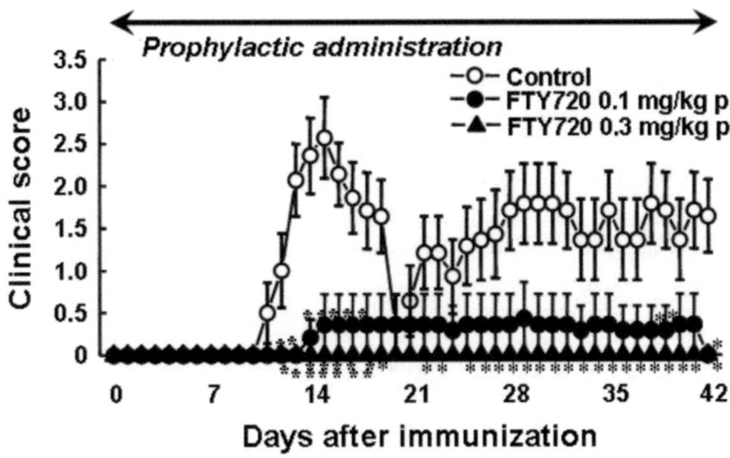

Fig. 5. Prophylactic effect of FTY720 on $PLP_{139-151}$-induced EAE in SJL/J mice. SJL/J mice were immunized with $PLP_{139-151}$ (50 µg/mouse) and Freund's complete adjuvant. FTY720 was administered orally to $PLP_{139-151}$-immunized SJL/J mice every day from the day of immunization for 42 days. Mice in the control groups were administered vehicle only. The results are expressed as the mean ± S.E.M. of 7 animals and statistical differences in clinical scores of EAE were calculated by Steel's test (*: $p<0.05$, **: $p<0.01$).

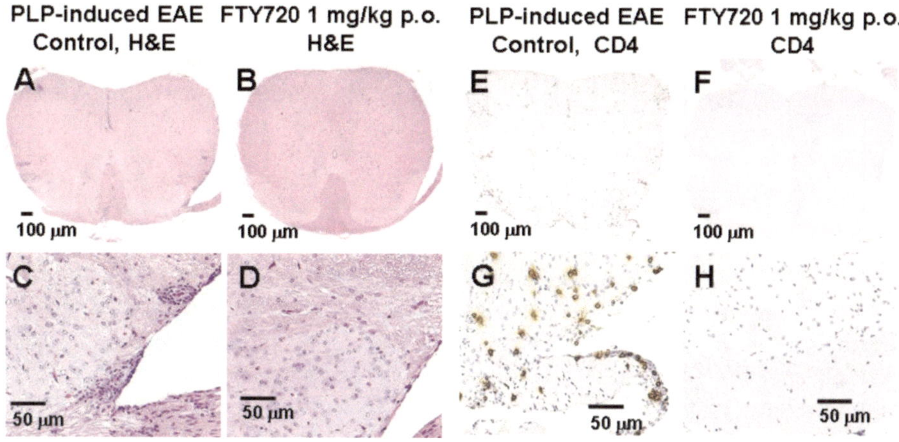

Fig. 6. Prophylactic administration of FTY720 decreased the infiltration of CD4+ T cells into the spinal cord of EAE mice induced by $PLP_{139-151}$ immunization. SJL/J mice were immunized with $PLP_{139-151}$ and administered FTY720 at 1 mg/kg p.o. prophylactically. On day 17 after immunization, the spinal cords of EAE-developed mice were obtained, and haematoxylin and eosin (H&E) and immunohistochemical staining with anti mouse CD4 monoclonal antibody (mAb) were performed.

5.2 Therapeutic effects of FTY720 on PLP-induced EAE in SJL/J mice

To evaluate the therapeutic effect of FTY720 on $PLP_{139-151}$-induced, relapsing EAE in SJL/J mice, EAE-developed mice were pooled, divided into 4 groups consisting of six mice, and administration of FY720 was started 17 days after immunization (Kataoka et al., 2005). EAE-associated clinical signs had decreased rapidly by day 21, and thereafter, relapse of EAE occurred in the vehicle-treated control group (Fig. 7A). By contrast, the relapse of EAE was markedly inhibited and no EAE-associated clinical signs were observed from day 32 to 59 in groups given FTY720 at 0.3 and 1 mg/kg therapeutically, indicating complete inhibition of EAE relapse. In the group given rm-IFN-β at 10000 IU three times a week intraperitoneally, the EAE score was significantly lowered at day 24, and relapse was delayed; however rm-IFN-β failed to inhibit the relapse of EAE. In another experiment, it was confirmed that the magnitude of the therapeutic effect of FTY720 is almost equal to that of prednisolone (1 mg/kg p.o.) (Fig. 7B). In the group given FTY720 at 1 mg/kg therapeutically, the area of demyelination and the infiltration of CD4+ T cells into the spinal cord were decreased as compared with the vehicle-treated control group (Fig. 8). Similarly, the relapse of EAE was markedly inhibited in groups given FTY720-P at 0.1 and 1 mg/kg therapeutically (Kataoka et al., 2005).

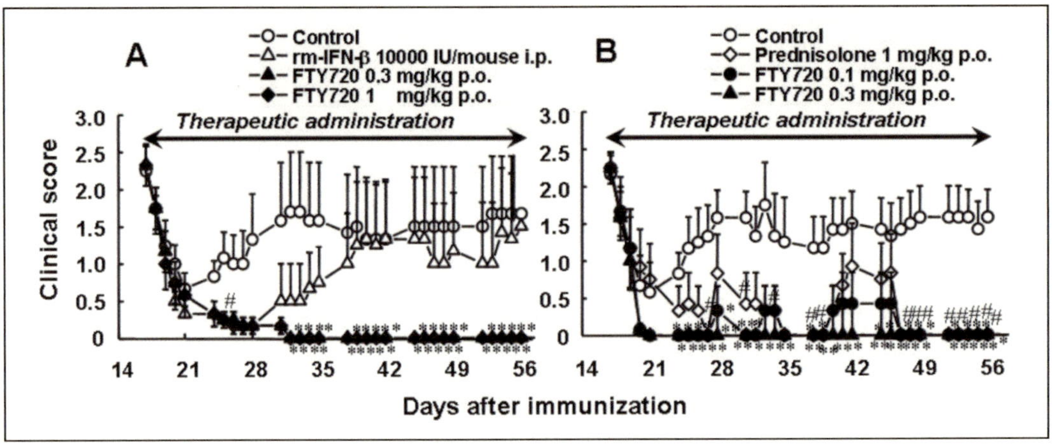

Fig. 7. Therapeutic effect of FTY720 on $PLP_{139-151}$-induced EAE in SJL/J mice. SJL/J mice were immunized with $PLP_{139-151}$ (50 μg/mouse) and Freund's complete adjuvant. EAE-developed mice were pooled, divided into 4 groups and administrations of FTY720 (daily), rm-IFN-β (3 times a week), and prednisolone (daily) were started from day 17. The results are expressed as the mean ± S.E.M of 7 animals and statistical differences in clinical scores of FTY720 groups were calculated by Steel's test (*: $p<0.05$, **: $p<0.01$), and those in rm-IFN-β or prednisolone were done by Mann Whitney *U* test (#: $p<0.05$).

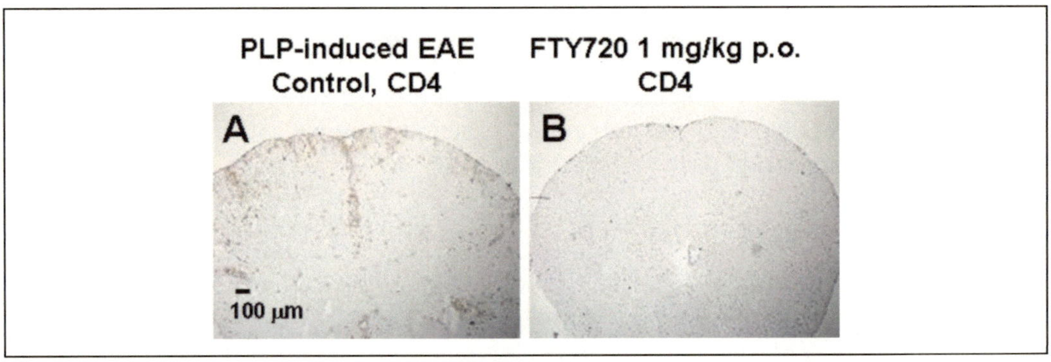

Fig. 8. Therapeutic administration of FTY720 decreased infiltration of $CD4^+$ T cells into spinal cords of $PLP_{139-151}$-induced EAE mice. SJL/J mice were immunized with $PLP_{139-151}$ and were administered FTY720 (1 mg/kg p.o.) therapeutically from day 17. The spinal cords of EAE-developed mice were obtained on day 28, and were performed immunohistochemical staining with anti-mouse CD4 mAb.

5.3 FTY720 shows a superior efficacy as compared with IFN-β in the relapse of EAE induced by PLP in SJL/J mice

To clarify the preventing effects of FTY720 on relapse of EAE induced by $PLP_{139-151}$ in SJL/J mice, EAE-developed mice were pooled on day 15 after immunization, divided into 5 groups, and administered vehicle or FTY720 (oral doses of 0.03, 0.1, 0.3 and 1 mg/kg) daily for 28 days (Fig. 9A) (Chiba et al., 2011). In vehicle-treated control group, the first phase of EAE was remitted with low EAE scores by day 8 after primary administration; however

EAE symptoms were relapsed on day 10 to day 14 spontaneously, reached to maximal level on day 16, and maintained with high severity thereafter. The relapse of EAE was significantly inhibited in all of FTY720 groups and no clinical symptoms were observed in the groups given FTY720 at 0.3 and 1 mg/kg from day 12 to day 28 after primary administration. As shown in Fig. 11B, all 12 mice in the control group experienced the relapse of EAE from day 6 to day 16 after primary administration. FTY720 (0.03 to 1 mg/kg) significantly prolonged the time to confirm the relapse of EAE and no relapse was seen in group given FTY720 at 0.3 and 1 mg/kg, indicating an almost complete prevention for the relapse of EAE (Fig. 9B).

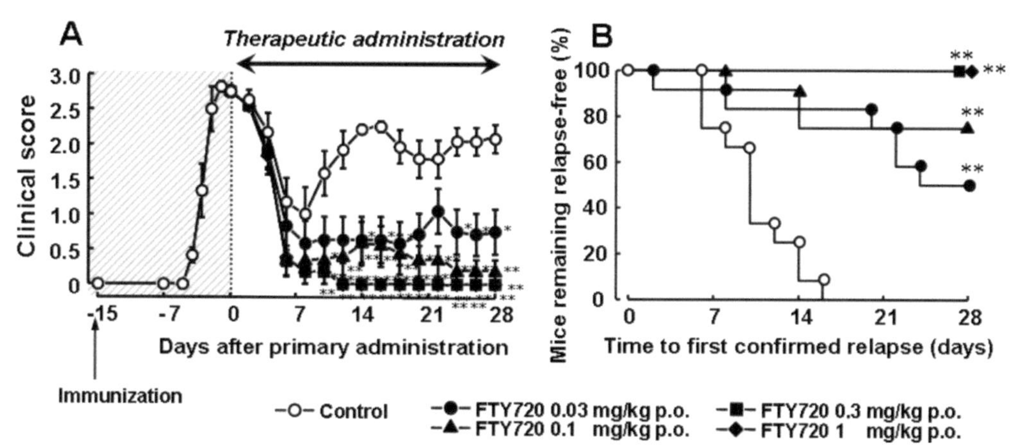

Fig. 9. Therapeutic effects of FTY720 on EAE induced by $PLP_{139-151}$ in SJL/J mice. SJL/J mice were immunized with $PLP_{139-151}$ (50 μg/mouse) and Freund's complete adjuvant. EAE-developed mice were divided into 5 groups on day 15 day after immunization and administered vehicle or FTY720 orally every day for 28 days. (A) Clinical scores are expressed as the mean ± S.E.M. of 12 mice. Statistical differences in EAE scores were calculated by Steel's test (*: $p<0.05$, **: $p<0.01$ versus vehicle-treated control group). (B) Results are expressed as the proportion of mice remaining relapse-free in total 12 mice. Statistical differences were calculated by generalized Wilcoxon test adjusted by Holm's multiple comparison test (**: $p<0.01$ versus vehicle-treated control group).

Next, we directly compared the therapeutic effects of FTY720 and rm-IFN-β on relapsing-remitting EAE induced by $PLP_{139-151}$ in SJL/J mice. EAE-developed SJL/J mice were divided into 5 groups on day 15 after immunization and then administered vehicle, FTY720 (0.1 and 0.3 mg/kg p.o., daily), or rm-IFN-β (3000 and 10000 IU/mouse, subcutaneously every other day) for 28 days (Chiba et al. 2011). The relapse of EAE was markedly inhibited in groups given FTY720 at 0.1 and 0.3 mg/kg (Fig. 10A). The administration of rm-IFN-β at low dose (3000 IU/mouse) resulted in only a slight but not significant inhibition on day 14 after primary administration. The high dose (10000 IU/mouse) of rm-IFN-β showed a significant reduction of EAE scores, suggesting a delay of the relapse in early period of administration; however even high dose of rm-IFN-β failed to prevent the relapse of EAE in latter period (Fig. 10B).

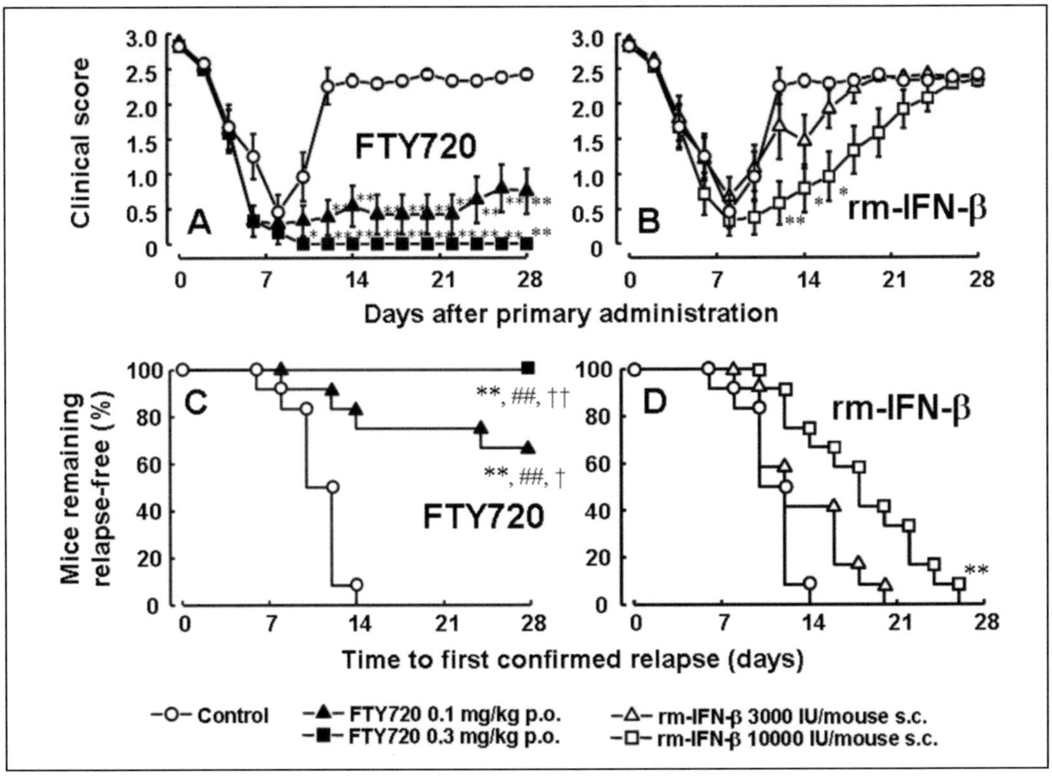

Fig. 10. Therapeutic effects of FTY720 and rm-IFN-β on relapsing remitting EAE induced by $PLP_{139-151}$ in SJL/J mice. EAE-developed mice were divided into 5 groups on day 15 after the immunization. Clinical scores are expressed as the mean ± S.E.M. of 12 mice. (A) Statistical differences in clinical scores of EAE were calculated by Steel's test (*: $p<0.05$, **: $p<0.01$ versus control). (B) Results are expressed as the proportion of mice remaining relapse-free in total 12 mice. Statistical differences were calculated by generalized Wilcoxon test adjusted by Holm's multiple comparison test (**: $p<0.01$ versus vehicle-treated control, ##: $p<0.01$ versus rm-IFN-β 3,000 IU/mouse, †: $p<0.05$, ††: $p<0.01$ versus rm-IFN-β 10,000 IU/mouse).

All mice in the control group experienced the relapse of EAE from day 6 to day 14. FTY720 at 0.1 and 0.3 mg/kg significantly prolonged the time to first confirmed relapse and no relapse was seen in 0.3 mg/kg group (Fig 10C). In the high dose of rm-IFN-β group, the relapse of EAE was significantly delayed; however all mice given rm-IFN-β experienced the relapse of EAE (Fig. 10D). Furthermore, FTY720 significantly prolonged the time to first confirmed relapse as compared with rm-IFN-β, indicating that FTY720 shows a superior efficacy for preventing the relapse of EAE as compared with rm-IFN-β.

5.4 FTY720 reduces infiltration of Th17 cells and Th1 cells into the spinal cords of PLP-induced EAE in SJL/J mice

To elucidate the effects of FTY720 on demyelination and infiltration of $CD4^+$ T cells into the CNS in EAE induced by $PLP_{139-151}$, the spinal cords were obtained from EAE mice given FTY720 or vehicle on day 28 after immunization.

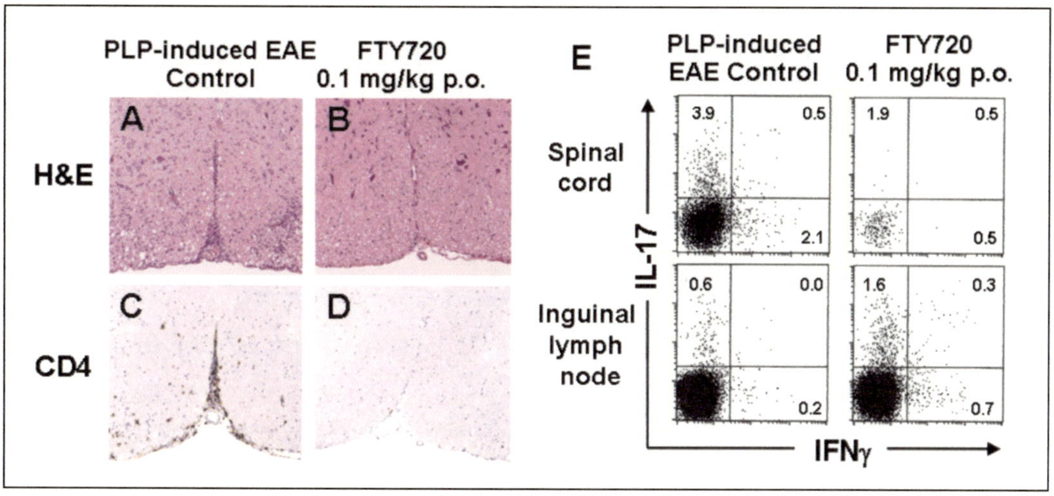

Fig. 11. Effects of FTY720 on the frequency of PLP-specific Th1 and Th17 cells in the spinal cords and inguinal lymph nodes in EAE mice immunized with $PLP_{139\text{-}151}$. SJL/J mice were immunized with $PLP_{139\text{-}151}$ (50 μg/mouse) and Freund's complete adjuvant. EAE-developed mice were administered FTY720 at an oral dose of 0.1 mg/kg therapeutically for 28 days. On the next day of the final administration, the spinal cords were obtained and were stained with H&E (A) control, (B) FTY720 0.1 mg/kg. Immunohistochemical staining of the spinal cords was performed by using anti-mouse CD4 mAb: (C) control, (D) FTY720 0.1 mg/kg. On the same day, lymphocytes were prepared from the spinal cords or inguinal lymph nodes of EAE mice and were cultured for 72 h in the presence of $PLP_{139\text{-}151}$ (50 μg/ml). After the culture, intracellular cytokine staining was performed by using anti-mouse CD4, anti-mouse IL-17, and anti-mouse IFN-γ mAbs (E).

Demyelination and infiltration of inflammatory cells including lymphocytes were observed in the spinal cord of EAE mice in vehicle-treated control group (Fig. 11A). Therapeutic oral administration of FTY720 at 0.1 mg/kg resulted in a marked reduction of demyelination and infiltration of inflammatory cells in the spinal cord of EAE mice (Fig. 11B). By immunohistochemical staining using anti-mouse CD4 mAb, it was revealed that $CD4^+$ T cells were infiltrated into the spinal cord, particularly the perivascular area and funiculus dorsalis in white matter under pia matter of EAE mice (Fig. 11C). Therapeutic administration of FTY720 at 0.1 mg/kg orally resulted in a marked reduction of infiltration of $CD4^+$ T cells into the spinal cord (Fig. 11D).

To clarify the involvement of myelin antigen-specific Th17 and Th1 cells, lymphocytes were prepared from the spinal cord and inguinal lymph nodes of EAE mice induced by immunization of $PLP_{139\text{-}151}$ on day 28 after immunization (Chiba et al., 2011). The obtained lymphocytes were re-stimulated with $PLP_{139\text{-}151}$ for 72 h *in vitro*. Then, the numbers of PLP-specific Th17 cells and Th1 cells were determined by intracellular cytokine staining using anti-mouse CD4, anti-IL-17, and anti-IFN-γ mAbs. Fig. 11E shows a typical pattern of intracellular cytokine staining of CD4-gated Th17 cells and Th1 cells in the spinal cord and inguinal lymph nodes in EAE mice. Significant numbers of Th17 and Th1 cells were found in the spinal cord from EAE mice, indicating the infiltration of myelin PLP-specific

Th17 and Th1 cells into the CNS of EAE mice. FTY720 (0.1 mg/kg) markedly reduced the infiltration of PLP-specific Th17 and Th1 cells into the spinal cord of EAE mice (Chiba et al., 2011). On the contrary, FTY720 increased the frequency of PLP-specific Th17 and Th1 cells in inguinal lymph nodes to approximately 3-fold, suggesting inhibition of egress of PLP-specific Th cells from draining lymph nodes. From these results, therapeutic effects of FTY720 on EAE are likely due to reduction of infiltration of myelin antigen-specific Th cells into the CNS.

6. Effects of FTY720 on MOG-induced EAE in C57BL/6 mice

6.1 Prophylactic and therapeutic effects of FTY720 on MOG-induced EAE in C57BL/6 mice

When C57BL/6 mice were immunized with MOG_{35-55} in the presence of Freund's complete adjuvant, chronic progressing EAE was developed on day 11 and reached a maximal level on day 17 after the immunization. Prophylactic administration of FTY720 at an oral dose of 0.3 mg/kg resulted in a marked delay in the onset of EAE and a significant reduction of EAE symptoms during the administration period (Fig. 12A) (Seki et al, 2010).

To examine the therapeutic effects of FTY720 on EAE induced by immunization with MOG_{35-55} in C57BL/6 mice, EAE-established mice were divided into 3 groups consisting of eleven mice 17 days after the immunization (Kataoka et al., 2005). When FTY720 (0.1 and 0.3 mg/kg) was administered therapeutically, the symptoms of EAE were significantly improved during the administration period (Fig. 12B).

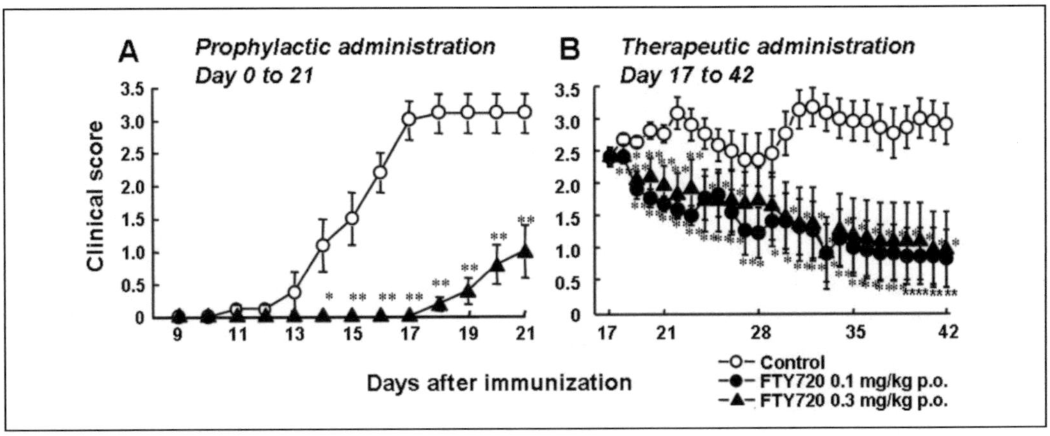

Fig. 12. Prophylactic and therapeutic effects of FTY720 on MOG_{35-55}-induced EAE in C57BL/6 mice. C57BL/6 mice were immunized with MOG_{35-55} (200 μg/mouse) in the presence of Freund's complete adjuvant. (A) Prophylactic administration of FTY720 was stared on the day of the immunization for 21 days. (B) EAE-developed mice were pooled, divided into 3 groups and therapeutic administration of FTY720 was started from day 17 for 25 days. Clinical scores are expressed as the mean ± S.E.M. of 10 to 11 mice. Statistical differences were calculated by Mann-Whitney *U* test (A) or Steel's test (B) (*: $p<0.05$, **: $p<0.01$).

To compare the therapeutic effects of FTY720 and rm-IFN-β on chronic EAE induced by MOG_{35-55} in C57BL/6 mice, EAE-developed mice were divided into 5 groups on day 15 after immunization with MOG_{35-55}, and administered vehicle or FTY720 (oral doses of 0.03, 0.1, 0.3 and 1 mg/kg) daily for 28 days (Chiba et al., 2011). Therapeutic administration of FTY720 at 0.1 to 1 mg/kg significantly improved the EAE-associated symptoms during administration period (Fig. 13A). On the other hand, rm-IFN-β (subcutaneous injection at 10,000 IU/mouse every other day) showed no clear effect on chronic EAE induced by MOG_{35-55} (Fig. 13B).

6.2 FTY720 reduces infiltration of Th17 and Th1 cells into the spinal cords of EAE mice induced by immunization with MOG

Demyelination and infiltration of inflammatory cells including lymphocytes were observed in the spinal cord of EAE mice in vehicle-treated control group (Fig. 14A). Therapeutic oral administration of FTY720 at 0.3 mg/kg resulted in a marked reduction of demyelination and infiltration of inflammatory cells in the spinal cord of EAE mice (Fig. 14B). By immunohistochemical staining using anti-mouse CD4 mAb, it was revealed that $CD4^+$ T cells were infiltrated into the spinal cord of EAE mice (Fig. 14C). Therapeutic administration of FTY720 at 0.1 mg/kg orally resulted in a marked reduction of infiltration of $CD4^+$ T cells into the spinal cord (Fig. 14D).

To examine the influence of FTY720 on infiltration of Th17 and Th1 cells into the CNS in MOG_{35-55}-induced EAE in C57BL/6 mice, the spinal cords were obtained from EAE mice

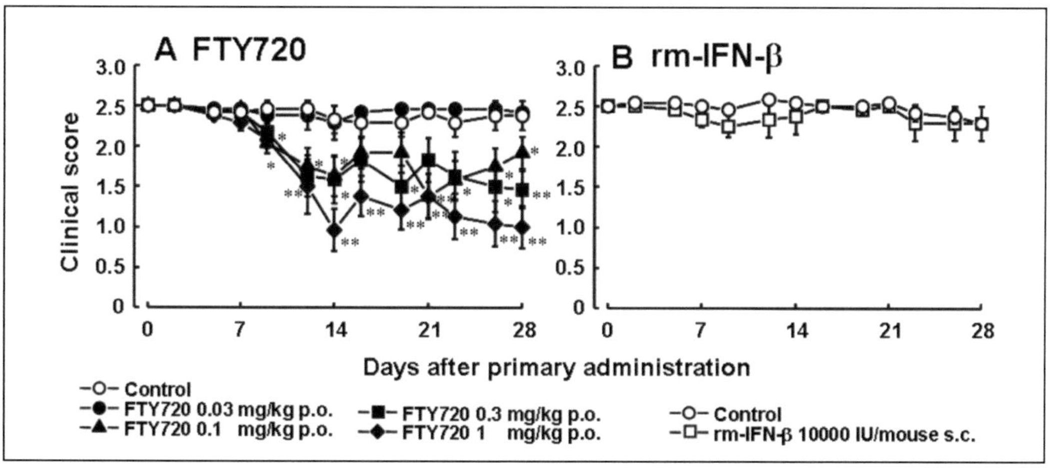

Fig. 13. Therapeutic effects of FTY720 and rm-IFN-β on chronic EAE induced by MOG_{35-55} in C57BL/6 mice. C57BL/6 mice were immunized with MOG_{35-55} (200 µg/mouse) and EAE-developed mice were divided into 5 groups on day 14 after immunization. (A) FTY720 was administered orally every day for 28 days. (B) rm-IFN-β was administered subcutaneously every other day for 28 days. Clinical scores are expressed as the mean ± S.E.M. of 12 mice. Statistical differences in EAE scores were calculated by Steel's test (*: $p<0.05$, **: $p<0.01$ versus control group). There was no significant difference between control group and rm-IFN-β-treated group by Wilcoxon test.

given FTY720 or vehicle for 16 days. Th17 and Th1 cells infiltrated into the spinal cords were determined by intracellular cytokine staining with mAbs against IL-17 and IFN-γ. Fig. 14E shows a typical pattern of intracellular cytokine staining of CD4-gated IL-17-expressing Th17 cells and IFN-γ-expressing Th1 cells in the lymphocytes infiltrated into the spinal cords in EAE mice (Seki et al., 2010). The number of $CD4^+$ T cells infiltrated into the spinal cords was markedly decreased to less than 10% of control EAE mice by prophylactic administration of FTY720 at 0.3 mg/kg for 16 days. Moreover, the frequency of Th17 and Th1 cells in the spinal cords of EAE mice given FTY720 was also markedly decreased compared with control (Fig. 14E). On the contrary, the frequency of Th17 and Th1 cells in draining inguinal lymph nodes was increased to approximately 2-fold by the administration of FTY720, suggesting that myelin antigen-specific Th cells can not exit from draining lymph nodes into periphery by treatment with FTY720.

FTY720-P, an active metabolite of FTY720, is known to show an agonistic activity on S1P receptors; however it acts as a functional antagonist of S1P1, because it strongly internalizes S1P1 receptor and reduces S1P responsiveness of lymphocytes (Brinkmann et al 2002, 2004; Chiba, 2005; Cyster, 2005; Maeda et al., 2007, Matloubian et al., 2004; Pham et al., 2008). As reported previously, mouse $CD4^+$ T cells shows a migratory respond to a physiological concentration (10 nM) of S1P and the S1P responsiveness of $CD4^+$ T cells is almost completely inhibited by pretreatment with 0.3 to 3 nM FTY720-P. To know whether myelin antigen-specific Th17 and Th1 cells generated in draining lymph nodes can respond a

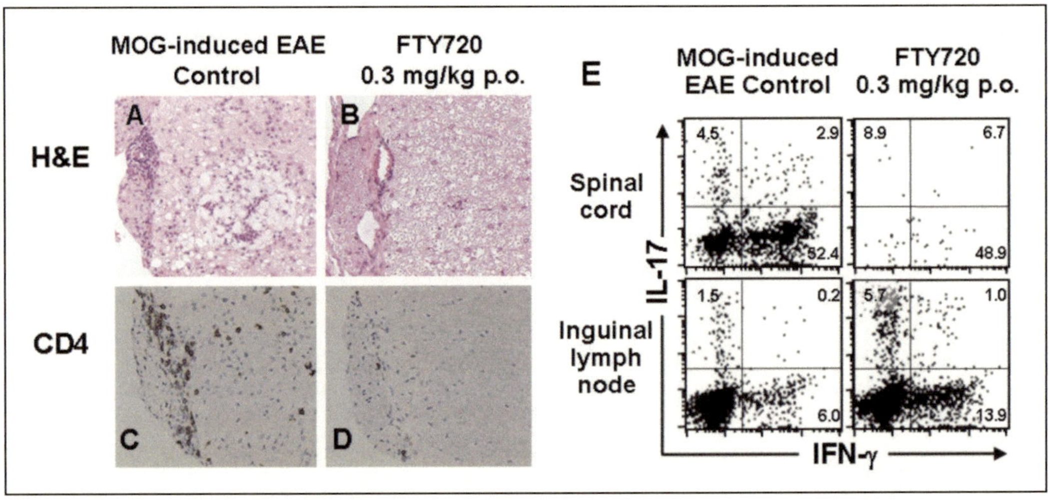

Fig. 14. Effects of FTY720 on the frequency of Th1 and Th17 cells in the spinal cords and inguinal lymph nodes in EAE mice immunized with MOG_{35-55}. C57BL/6 mice were immunized with P MOG_{35-55} (200 µg/mouse) and EAE-developed mice were administered FTY720 at an oral dose of 0.1 mg/kg therapeutically for 28 days. On the next day of the final administration, the spinal cords were obtained and were stained with H&E (A) control, (B) FTY720 0.3 mg/kg. Immunohistochemical staining of the spinal cords was performed by using anti-mouse CD4 mAb: (C) control, (D) FTY720 0.3 mg/kg. On the same day, lymphocytes were prepared from the spinal cords or inguinal lymph nodes of EAE mice and intracellular cytokine staining was performed by using anti-mouse CD4, anti-mouse IL-17, and anti-mouse IFN-γ mAbs (E).

physiological concentration of S1P, we performed migration assays toward S1P using lymphocytes from draining inguinal lymph nodes of EAE-established mice. Both Th17 and Th1 cells prepared from the draining lymph nodes of EAE mice could migrate toward 10 nM S1P and the migratory response of these Th cells toward S1P was almost completely inhibited by pretreatment with 1 nM of FTY720-P (Seki et al., 2010). On the other hand, FTY720-P up to 100 nM showed no clear effect on the generation of either Th17 or Th1 cells from naïve $CD4^+$ T cells *in vitro* (Seki et al., 2010). Moreover FTY720-P did not affect IL-17 production from Th17 cells or IFN-γ production from Th1 cells *in vitro* (Seki et al., 2010). From these results, it is suggested that FTY720-P converted from FTY720 inhibits egress of myelin antigen-specific Th17 and Th1 cells from draining lymph nodes, reduces infiltration of these Th cells into the CNS, and shows ameliorating effects on EAE in mice.

7. Therapeutic effects of FTY720 in relapsing remitting MS

The first clinical evidence that FTY720 has the therapeutic benefits in MS was provided in a 6-month, placebo-controlled Phase II trial involving 281 patients with relapsing remitting MS (RRMS) (Kappos et al., 2006). Patients receiving FTY720 at an oral dose of 1.25 mg or 5.0 mg daily had a significant lower median total number of gadolinium-enhancing lesions (the primary end point) on magnetic resonance imaging (MRI) than those receiving placebo. The annualized relapse rates in groups given 1.25 mg and 5.0 mg of FTY720 were 0.35 and 0.36, respectively and were significantly lower than that in the placebo group (0.77). By extension study for additional 6 months, the number of gadolinium-enhanced lesions and relapse rates remained low in groups given FTY720 and both measures decreased in patients who switched from placebo to FTY720.

In FTY720-treated MS patients, the number of IL-17-expressing $CD4^+$ T cells were reduced by >95% in the peripheral blood suggesting that FTY720 inhibits egress of Th17 cells from the SLO and reduces the infiltration of Th17 cells into the CNS (Brinkman, 2009; Mehling et al., 2010). In addition, FTY720 primarily reduced the numbers of $CCR7^+$ $CD45RA^+$ naïve T cells and $CCR7^+$ $CD45RA^-$ central memory T cells in the blood in MS patients, because these T cells express the homing receptor CCR7, recirculate through the lymph nodes, and can be sequestered into the lymph nodes by FTY720. In contrast, $CCR7^-$ $CD45RA^-$ and $CCR7^-$ $CD45RA^+$ effector memory T cell subsets are not sequestered into the SLO and are remained in the blood when FTY720 is administered. These results suggest that FTY720 effectively inhibits infiltration of pathogenic $CD4^+$ T cells including Th 17 cells into the CNS in MS patients whereas FTY720 does not affect the function of effector memory T cells that play an important role in the prevention of infection.

FTY720 was evaluated in a 24-month, double blind Phase III study (FREEDOMS study), involving 1272 patients RRMS (Kappos et al., 2010). The patients were randomized to receive a daily oral dose of FTY720 at 0.5 mg or 1.25 mg, or placebo. The annualized relapse rates in groups given 0.5 mg and 1.25 mg of FTY720 were 0.18 and 0.16, respectively and were significantly lower than that in the placebo group (0.40). FTY720 at 0.5 mg and 1.25 mg significantly reduced the risk of disability progression over 24-month period. The cumulative probability of disability progression confirmed after 3 months was 17.7% with 0.5 mg FTY720, 16.6% with 1.25 mg FTY720, and 24.1% with placebo. FTY720 at 0.5 mg and 1.25 mg showed improved effects compared with placebo with regard to the MRI-related measures (number of new or enlarged lesions on T2-weightend images, gadolinium-enhanced lesions, and brain-volume loss).

FTY720 was also evaluated in a 12-month, double blind, double dummy Phase III study (TRANSFORMS study) involving 1292 patients with relapsing remitting MS, comparing FTY720 with IFN-β-1a (AVONEX®), an established therapy for MS (Cohen et al., 2010). Patients were randomized to receive a daily dose of 0.5 mg or 1.25 mg FTY720 orally, or a weekly intramuscular injection of IFN-β-1a. The annualized relapse rates in groups given FTY720 0.5 mg and 1.25 mg were 0.16 and 0.20 respectively, and were significantly lower than that in the group receiving IFN-β-1a (0.33). FTY720 at 0.5 mg and 1.25 mg showed improved effects compared with IFN-β-1a with regard to MRI-related measures. These Phase III studies demonstrated that oral FTY720 had superior efficacy compared with intramuscular IFN-β-1a and placebo with regard to reducing the rate of relapse and MRI-related measures of inflammatory lesion activity. Based on these results from clinical trials, FTY720 was approved by the United States Food and Drug Administration in September 2010 as a first-line treatment for RRMS. As FTY720 has been approved more than 40 countries including EU at this time, it is presumed that oral FTY720 provides a new therapeutic approach for RRMS.

8. Conclusion

FTY720, a S1P receptor modulator, acts as a functional antagonist at S1P1 and is highly effective in EAE in rats and mice. We directly compared the therapeutic effects of FTY720 and rm-IFN-β on relapse and progression of EAE in mice. When FTY720 at oral doses of 0.03 to 1 mg/kg was administered daily after establishment of EAE induced by myelin PLP in SJL/J mice, relapse of EAE was significantly inhibited during administration period. Subcutaneous injection of rm-IFN-β (10,000 IU/mouse) also inhibited the relapse of EAE at early period; however EAE was relapsed in all the mice within administration period. Therapeutic administration of FTY720 (0.03 to 1 mg/kg) significantly improved the symptoms of chronic EAE induced by myelin oligodendrocyte glycoprotein in C57BL/6 mice whereas rm-IFN-β (10,000 IU/mouse) showed no clear effect. These results indicate that FTY720 is more efficacious in mouse EAE as compared with rm-IFN-β. FTY720 markedly reduced the frequency of myelin antigen-specific Th17 and Th1 cells in the spinal cord of EAE mice, suggesting that the ameliorating effects of FTY720 on EAE are likely due to reduction of infiltration of myelin antigen-specific Th17 and Th1 cells into the CNS. Recently, it has been strongly suggested that FTY720-P directly acts as a functional antagonist at S1P1 on not only lymphocytes but also neural cells, particularly astrocytes because the therapeutic effects of FTY720 on EAE was lost in CNS mutants lacking S1P1 on astrocytes but not neuron. Since FTY720 possesses a completely new mechanism of action and shows superior efficacy compared with intramuscular IFN-β-1a (AVONEX®) in relapsing remitting MS patients, FTY720 should be a useful therapeutic approach for MS.

9. Acknowledgment

The authors thank Mamoru Koyama, Kyoko Shimano, Mikako Murase, Sachiko Mochiduki, Dr. Yasuhiro Maeda, Dr. Atsushi Fukunari, and Dr. Kunitomo Adachi, Mitsubish Tanabe Pharma Corporation, for their fruitful collaborations.

10. References

Adachi, K.; Kohara, T.; Nakao, N.; Arita, M.; Chiba, K.; Mishina, T.; Sasaki, S. & Fujita, T. (1995). Design, synthesis and structure-activity relationships of 2-substituted-2-amino-1,3-propanediols: discovery of a novel immunosuppressant, FTY720. *BioMed. Chem. Lett.*, Vol. 5, pp. 853-856

Anderson, S. A.; Shukaliak-Quandt, J.; Jordan, E. K. et al. (2004). Magnetic resonance imaging of labeled T-cells in a mouse model of multiple sclerosis. *Ann. Neurol.*, Vol. 55, pp. 654-659

Balatoni, B.; Storch, MK.; Swoboda, E. M.; Schonborn, V.; Koziel, A.; Lambrou, G. N. et al. (2007). FTY720 sustains and restores neuronal function in the DA rat model of MOG-induced experimental autoimmune encephalomyelitis. *Brain Res. Bull.*, Vol. 74, pp. 307-316

Bettelli, E.; Carrier, Y.; Gao, W.; Korn, T.; Strom, T. B.; Oukka, M. et al. (2006). Reciprocal developmental pathways for the generation of pathogenic effector TH17 and regulatory T cells. *Nature*, Vol. 441, pp. 235-238.

Billich, A.; Bornancin, F.; Devay, P.; Mechtcheriakova, D.; Urtz, N. & Baumruker, T. (2003). Phosphorylation of the immunomodulatory drug FTY720 by sphingosine kinases. *J. Biol. Chem.* Vol. 278, pp. 47408-47415

Bright, J. J.; Du, C.; Coon, M.; Sriram, S. & Klaus, S. J. (1998). Prevention of experimental allergic encephalomyelitis via inhibition of IL-12 sgnaling and IL-12-mediated Th1 differentiation: an effect of the novel anti-inflammatory drug lisofylline. *J. Immunol.*, Vol. 161, pp. 7015-7022.

Brinkmann, V.; Pinschewer, D.; Chiba, K. & Feng, L. (2000). FTY720: a novel transplantation drug that modulates lymphocyte traffic rather than activation. *Trends in Pharmacological Sciences*, Vol. 21, pp. 49-52

Brinkmann, V.; Davis, M. D.; Heise, C. E.; Albert, R.; Cottens, S.; Hof, R.; Bruns, C.; Prieschl, E.; Baumruker, T.; Hiestand, P.; Foster, C. A.; Zollinger, M. & Lynch, K. R. (2002a). The immune modulator FTY720 targets sphingosine 1-phosphate receptors. *J. Biol. Chem.*, Vol. 277, pp. 21453-21457

Brinkmann, V. & Lynch, K. R. (2002b). FTY720: target G-protein-coupled receptors for sphingosine 1-phosphate in transplantation and autoimmunity. *Curr. Opin. Immunol.*, Vol. 14, pp. 569-575

Brinkmann, V.; Cyster, J. G. & Hla, T. (2004). FTY720: sphingosine 1-phosphate receptor-1 in the control of lymphocyte egress and endothelial barrier function. *Am. J. Transplant.* Vol. 4, pp. 1019-1025

Brinkmann, V. (2009). FTY720 (fingolimod) in Multiple Sclerosis: therapeutic effects in the immune and the central nervous system. *Br. J. Pharmacol.*, Vol. 158, pp 1173-1182

Chiba, K.; Hoshino, Y.; Suzuki, C.; Masubuchi, Y.; Yanagawa, Y.; Ohtsuki, M.; Sasaki, S. & Fujita, T. (1996). FTY720, a novel immunosuppressant possessing unique mechanisms I. Prolongation of skin allograft survival and synergistic effect in combination with cyclosporin A in rats. *Transplant. Proc.*, Vol. 28, pp. 1056-1059

Chiba, K. & Adachi, K. (1997). FTY720, immunosuppressant. *Drugs of the Future*, Vol. 22, pp. 18-22

Chiba, K.; Yanagawa, Y.; Masubuchi, Y.; Kataoka, H.; Kawaguchi, T.; Ohtsuki, M. & Hoshino, Y. (1998). FTY720, a novel immunosuppressant, induces sequestration of circulating mature-lymphocytes by acceleration of lymphocyte homing in rats. I.

FTY720 selectively decreases the number of circulating mature lymphocytes by acceleration of lymphocyte homing. *J. Immunol.*, Vol. 160, pp. 5037-5044

Chiba, K.; Yanagawa, Y.; Kataoka, H.; Kawaguchi, T.; Ohtsuki, M. & Hoshino, Y. (1999). FTY720, a novel immunosuppressant, induces sequestration of circulating lymphocytes by acceleration of lymphocyte homing. *Transplant. Proc.*, Vol. 31, pp. 1230-1233

Chiba, K.; Hoshino, Y.; Ohtsuki, M.; Kataoka, H.; Maeda, Y.; Matsuyuki, K.; Sugahara, K.; Kiuchi, M.; Hirose, R. & Adachi, K. (2005). Imunosuppressive activity of FTY720, sphingosine 1-phosphate receptor agonist: I. Prevention of allograft rejection in rats and dogs by FTY720 and FTY720-phosphate. *Transplant. Proc.*, Vol. 37, pp. 102-106

Chiba, K. (2005). FTY720, a new class of immunomodulator, inhibits lymphocyte egress from secondary lymphoid tissues and thymus by agonistic activity at sphingosine 1-phosphate receptors. *Pharmacol. Ther.*, Vol. 108, pp. 308-319

Chiba, K.; Matsuyuki, H.; Maeda, Y. & Sugahara, K. (2006). Role of sphingosine 1-phosphate receptor type 1 in lymphocyte egress from secondary lymphoid tissues and thymus. *Cell. Mol. Immunol.*, Vol. 3, pp. 11-19

Chiba, K. (2010). Sphingosine 1-phosphate receptor type 1 as a novel target for the therapy of autoimmune diseases. *Inflamm. Regen.* Vol. 30, pp. 160-168

Chiba, K.; Kataoka, H.; Seki, N.; Shimano, K.; Koyama, M.; Fukunari, A.; Sugahara, K. & Sugita T. (2011). Fingolimod (FTY720), sphingosine 1-phosphate receptor modulator, shows superior efficacy as compared with interferon-β in mouse experimental autoimmune encephalomyelitis. *Int. Immunopharmacol.*, Vol. 11, pp. 366-372

Choi, J. W.; Gardell, S. E.; Herr, D. R.; Rivera, R.; Lee, C.-W.; Noguchi, K. et al. (2011). FTY720 (fingolimod) efficacy in an animal model of multiple sclerosis requires astrocyte sphingosine 1-phosphate receptor 1 (S1P1) modulation. *Proc. Natl. Acad. Sci. USA.*, Vol. 108, pp. 751-756

Cohen, J. A.; Barkhof, F.; Comi, G.; Hartung, H. P.; Khatri, B. O.; Montalban, X. et al. (2010). Oral fingolimod or intramuscular interferon for relapsing multiple sclerosis. *N. Engl. J. Med.*, Vol. 362, pp. 402-415

Cyster, J. G. (2005). Chemokines, sphingosine-1-phosphate, and cell migration in secondary lymphoid organs. *Annu. Rev. Immunol.* Vol. 23, pp. 127-159.

Foster, C. A.; Howard, L. M.; Schweitzer, A.; Persohn, E. Hiestand, P. C.; Balatoni, B. et al. (2007). Brain penetration of the oral immunomodulatory drug FTY720 and its phosphorylation in the central nervous system during experimental autoimmune encephalomyelitis: consequences for mode of action in multiple sclerosis. *J. Pharmacol. Exp. Ther.* Vol. 323, pp. 469-475

Fujino, M.; Funeshima, N.; Kitazawa, Y.; Kimura, H.; Amemiya, H.; Suzuki, S. & Li, X. K. (2003). Amelioration of experimental autoimmune encephalomyelitis in Lewis rats by FTY720 treatment. *J. Pharmacol. Exp. Ther.*, Vol. 305, pp. 70-77

Fujita, T.; Inoue, K.; Yamamoto. S.; Ikumoto, T.; Sasaki, S.; Toyama, R.; Chiba, K.; Hoshino. Y. & Okumoto, T. (1994a). Fungal metabolites. Part 11. A potent immunosuppressive activity found in Isaria sinclairii metabolite. *J. Antibiotics*, Vol. 47, pp. 208-215

Fujita, T.; Inoue, K., Yamamoto, S.; Ikumoto, T.; Sasaki, S.; Toyama, R.; Chiba, K.; Hoshino, Y. & Okumoto, T. (1994b). Fungal metabolites. Part 12. Potent immunosuppressant,

14-deoxomyruiocin, (2*S*,3*R*,4*R*)-(*E*)-2-amino-3,4-dihydroxy-2-hydroxymethyleicos-6-enoic acid and structure-activity relationships of myriocin derivatives. *J. Antibiotics*, Vol. 47, pp. 216-224

Fujita, T.; Yoneta, M.; Hirose, R.; Sasaki, S.; Inoue, K.; Kiuchi, M.; Hirase, S.; Adachi, K.; Arita, M. & Chiba, K. (1995). Simple compounds, 2-alkyl-2-amino-1,3-propane-diols have potent immunosuppressive activity. *BioMed. Chem. Lett.*, Vol. 5, pp. 847-852

Fujita, T.; Hirose, R.; Yoneta, M.; Sasaki, S.; Inoue, K.; Kiuchi, M.; Hirase, S.; Chiba, K.; Sakamoto, H. & Arita, M. (1996). Potent immunosuppressants, 2-alkyl-2-aminopropane-1,3-diols. *J. Med. Chem.*, Vol. 39, pp. 4451-4459

Goodkin, D. E.; Reingold, S.; Sibley, W. et al. (1999). Guide lines for clinical trials for new therapeutic agents in multiple sclerosis: reporting extended results from phase III clinical trials-National Multiple Sclerosis Society Advisory Committee on Clinical Trials of New Agents in Multiple sclerosis. *Ann. Neurol.*, Vol. 46, pp. 132-134

Graler, M. H. & Goetzl, E. J. (2004). The immunosuppressant FTY720 down-regulates sphingosine 1-phosphate G-protein-coupled receptors. *FASEB J.* Vol. 18, pp. 551-553.

Hla, T.; Lee, M. J.; Ancellin, N.; Paik, J. H. & Kluk, M. J. (2001). Lysophospholipids - receptor revelations. *Science*, Vol. 294, pp. 1875-1878

Hoshino, Y.; Suzuki, C.; Ohtsuki, M.; Masubuchi, Y.; Amano, Y. & Chiba, K. (1996). FTY720, a novel immunosuppressant possessing unique mechanisms II. Long-term graft survival induction in rat hetrotopic cardiac allograft and synergistic effect in combination with cyclosporin A. *Transplant. Proc.*, Vol. 28, pp. 1060-1061

Kataoka, H.; Sugahara, K.; Shimano, K,; Teshima, K.; Koyama, M.; Fukunari, A. & Chiba, K. (2005). FTY720, sphingosine 1-phosphate receptor modulator, ameliorates experimental autoimmune encephalomyelitis by inhibition of T cell infiltration. *Cell. Mol. Immunol.* Vol. 2, pp. 439-448

Kappos, L.; Antel, J.; Comi, G.; Montalban, X.; O'Connor, P.; Polman, C. H. et al. (2006). Oral fingolimod (FTY720) for relapsing multiple sclerosis. *N. Engl. J. Med.*, Vol. 355, pp. 1124-1140

Kappos, L.; Radue, E. W.; O'Connor, P.; Polman, C.; Hohlfeld, R.; Calabresi, P. et al. (2010). A placebo-controlled trial of oral fingolimod in relapsing multiple sclerosis. *N. Engl. J. Med.*, Vol. 362, pp. 387-401

Kawaguchi, T.; Hoshino, Y.; Rahman, F.; Amano, Y.; Higashi, H.; Kataoka, H.; Ohtsuki, M.; Teshima, K.; Chiba. K.; Kakefuda, T. & Suzuki, S. (1996). FTY720, a novel immunosuppressant possessing unique mechanisms III. Synergistic prolongation of canine renal allograft survival in combination with cyclosporin A. *Transplant. Proc.*, Vol. 28, pp. 1062-1063

Kimura, Y.; Sato, K.; Kuwabara, A.; Tomura, H.; Ishikawa, M.; Kobayashi, I.; Ui, M. & Okajima, F. (2001). Sphingosine 1-phosphate may be a major component of plasma lipoproteins responsible for the cytoprotective actions in human umbilical vein endothelial cells. *J. Biol. Chem.* Vol. 276, pp. 31780-31785

Kiuchi, M.; Adachi, K.; Kohara, T.; Minoguchi, M.; Hanano, T.; Aoki, Y.; Mishina, T.; Arita, M.; Nakao, N.; Ohtsuki, M.; Hoshino, Y.; Teshima, K.; Chiba, K.; Sasaki, S. & Fujita, T. (2000). Synthesis and immunosuppressive activity of 2-substituted 2-aminopropane-1,3-diols and 2-aminoethanols. *J. Med. Chem.*, Vol. 43, pp. 2946-2961

Kiuchi, M.; Adachi, K.; Tomatsu, A.; Chino, M.; Takeda, S.; Tanaka, Y.; Maeda, Y.; Sato, N.; Mitsutomi, N.; Sugahara, K. & Chiba, K. (2005). Asymmetric synthesis and biological evaluation of the enantiomeric isomer of immunosuppressive FTY720-phosphate. *Bioorg. Med. Chem.*, Vol. 13, pp 425-432

Komiyama, Y.; Nakae, S.; Matsuki, T.; Nambu, A.; Ishigame, H.; Kakuta, S. et al. (2006). IL-17 plays an important role in the development of experimental autoimmune encephalomyelitis. *J. Immunol.*, Vol. 177, pp. 566-573

Kroenke, M. A.; Carlson, T. J.; Andjelkovic, A. V. & Segal, B. M. (2008). IL-12- and IL-23-modulated T cells induce distinct types of EAE based on histology, CNS chemokine profile, and response to cytokine inhibition. *J. Exp. Med.*, Vol. 205, pp. 1535-1541

Kuchroo, V. K.; Martin, C. A.; Greer, J. M.; Ju, S. T.; Sobel, R. A. & Dorf, M. E. (1993). Cytokines and adhesion molecules contribute to the ability of myelin proteolipid protein-specific T cell clones to mediate experimental allergic encephalomyelitis. *J. Immunol.*, Vol. 151, pp. 4371-4382

Langrish, C. L.; Chen, Y.; Blumenschein, W. M.; Mattson, J.; Basham, B.; Sedgwick, J. D. et al. (2005). IL-23 drives a pathogenic T cell population that induces autoimmune inflammation. *J. Exp. Med.*, Vol. 201, pp. 233-240

Li, H.; Meno-Tetang, G. L.; Chiba, K.; Arima, N.; Heinig, P. & Jusko, W. L. (2002). Pharmacokinetics and cell trafficking dynamics of 2-amino-2-[2-(4-octylphenyl) ethyl]propane-1,3-diol hydrochloride (FTY720) in cynomolgus monkeys after single oral and intravenous doses. *J. Pharmacol. Exp. Ther.*, Vol. 301, pp. 519-526

Lo, C. G.; Xu, Y.; Proia, R. & Cyster, J. G. (2005). Cyclical modulation of sphingosine-1-phosphate receptor 1 surface expression during lymphocyte recirculation and relationship to lymphoid organ transit. *J. Exp. Med.*, Vol. 201, pp. 291-301

Lublin, F. D. & Reingold, S. C. (1996). Defining the clinical course of multiple sclerosis: results of an international survey. National Multiple Sclerosis Society (USA) Advisory Committee on Clinical Trials of New Agents in Multiple Sclerosis. *Neurology*, Vol. 46, pp. 907-911.

Maeda, Y.; Matsuyuki, H.; Shimano, K.; Kataoka, H.; Sugahara, K. & Chiba, K. (2007). Migration of CD4 T cells and dendritic cells toward sphingosine 1-phosphate (S1P) is mediated by different receptor subtypes: S1P regulates the functions of murine mature dendritic cells via S1P receptor type 3. *J. Immunol.*, Vol. 178, pp. 3437-3446

Mandala, S.; Hajdu, R.; Bergstrom, J.; Quackenbush, E.; Xie, J.; Milligan, J.; Thornton, R.; Shei, G.-J.; Card, D.; Keohane, C.; Rosenbach, M.; Hale, J.; Lynch, C. L.; Rupprecht, K.; Parsons, W. & Rosen, H. (2002). Alteration of lymphocyte trafficking by sphingosine-1-phosphate receptor agonists. *Science*, Vol. 296, pp. 346-349

Martin, R.; McFarland, H. F. & McFarlin, D. E. (1992). Immunological aspects of demyelinating diseases. *Annu. Rev. Immunol.*, Vol. 10, pp. 153-187

Martin, R. & McFarland, H. F. (1995). Immunological aspects of experimental allergic encephalomyelitis and multiple sclerosis. *Crit. Rev. Clin. Lab. Sci.*, Vol. 32, pp. 121-182

Masubuchi, Y.; Kawaguchi, T.; Ohtsuki, M.; Suzuki, C.; Amano, Y.; Hoshino, Y. & Chiba, K. (1996). FTY720, a novel immunosuppressant possessing unique mechanisms IV. Prevention of graft versus host reactions in rats. *Transplant. Proc.*, Vol. 28, pp. 1064-1065

Matloubian, M.; Lo, C. G.; Cinamon, G.; Lesneski, M. J.; Xu, Y.; Brinkmann, V.; Allende, M.; Proia, R. & Cyster, J. G. (2004). Lymphocyte egress from thymus and peripheral lymphoid organs is dependent on S1P receptor 1. *Nature*, Vol. 427, pp. 355-360

Mehling, M.; Lindberg, R.; Raulf, F.; Kuhle, J.; Hess, C., Kappos, L. & Brinkmann, V. (2010) Th17 central memory T cells are reduced by FTY720 in patients with multiple sclerosis. *Neurology*, Vol. 75, pp. 403-410.

Merrill, J. E., Kono, D. H.; Clayton, J.; Ando, D. G.; Hinton, D. R. & Hofman, F. M. (1992). Inflammatory leukocytes and cytokines in peptide-induced disease of experimental allergic encephalomyelitis in SJL and B10.PL mice. *Proc. Natl. Acad. Sci. USA.*, Vol. 89, pp. 574-578

Owens, T. & Sriram, S. (1995). The immunology of multiple sclerosis and its animal model, experimental allergic encephalomyelitis. *Neurol. Clin.*, Vol. 13, pp 51-73

Paugh, S. W.; Payne, S. G.; Barbour, S. E.; Milstien, S. & Spiegel, S. (2003). The immunosuppressant FTY720 is phosphorylated by sphingosine kinase type 2. *FEBS Lett.*, Vol. 554, pp. 189-193

Pham, T. H.; Okada, T.; Matloubian, M.; Lo, C. G. & Cyster, J. G. (2008). S1P1 receptor signalling overrides retention mediated by G alpha i-coupled receptors to promote T cell egress. *Immunity*, Vol. 28, pp. 122-133

Pyne, S. & Pyne, N. (2000). Sphingosine 1 phosphate signalling via the endothelial differentiation gene family of G-protein-coupled receptors. *Pharmacol. Ther.* Vol. 88, pp. 115-131

Rammohan, K. W. (2003). Axonal injury in multiple sclerosis. *Curr. Neurol. Neurosci. Rep.*, Vol. 3, pp. 231-237.

Sasaki, S.; Hashimoto, R.; Kiuchi, M.; Inoue, K.; Ikumoto, T.; Hirose, R.; Chiba, K.; Hoshino, Y. & Okumoto, T. (1994). Fungal metabolites. Part 14. Novel potent immunosuppressants, Mycestericins, produced by *Mycelia sterilia*. *J. Antibiotics*, Vol. 47, pp. 420-433

Seamons, A.; Perchellet, A. & Goverman, J. (2003). Immune tolerance to myelin proteins. *Immunol. Rev.*, Vol. 28, pp. 201-221

Seki, N.; Maeda, Y.; Kataoka, H.; Sugahara, K.; Sugita, T. & Chiba, K. (2010). Fingolimod (FTY720) ameliorates experimental autoimmune encephalomyelitis (EAE): II. FTY720 decreases infiltration of Th17 and Th1 cells into the central nervous system in EAE. *Inflamm. Regen.* Vol. 30, pp. 545-551

Steinman, L. (2010). Mixed results with modulation of TH-17 cells in human autoimmune diseases. *Nat. Immunol.*, Vol. 11, pp. 41-44

Stromnes, I. M.; Cerretti, L. M.; Liggitt, D.; Harris, R. A. & Goverman, J. M. (2008). Differential regulation of central nervous system autoimmunity by T_H1 and T_H17 cells. *Nat. Med.*, Vol. 14, pp. 337-342

Suzuki, S.; Enosawa, S.; Kakefuda, T.; Shinomiya, T.; Amari, M.; Naoe, S.; Hoshino, Y. & Chiba, K. (1996) A novel immunosuppressant, FTY720, having an unique mechanism of action induces long-term graft acceptance in rat and dog allotransplantation. *Transplantation*, Vol. 61, pp. 200-205

Suzuki, S.; Kakefuda, T.; Amemiya, H.; Chiba, K.; Hoshino, Y.; Kawaguchi, T.; Kataoka, H. & Rahman, F. (1998). An immunosuppressive regime using FTY720 combined with cyclosporin in canine kidney transplantation. *Transpl. Int.*, Vol. 11, pp. 95-101

Webb, M.; Tham, C. S.; Lin, F. F.; Lariosa-Willingham, K.; Yu, N.; Hale, J. et al. (2004). Sphingosine 1-phosphate receptor agonists attenuate relapsing-remitting experimental autoimmune encephalitis in SJL mice. *J. Neuroimmunol.*, Vol. 153, pp. 108-121

Windhagen, A.; Newcombe, J.; Dangond, F. et al. (1995). Expression of costimulatory molecules B7-1 (CD80), B7-2 (CD86), and interleukin 12 cytokines in multiple sclerosis lesions. *J. Exp. Med.*, Vol. 182, pp. 1985-1996

Yagi, H.; Kamba, R.; Chiba, K.; Soga, H.; Yaguchi, K.; Nakamura, M. & Itoh, T. (2000). Immunosuppressant FTY720 inhibits thymocyte emigration. *Eur. J. Immunol.*, Vol. 30, pp. 1435-1444

Yanagawa, Y.; Sugahara, K.; Kataoka, H.; Kawaguchi, T.; Masubuchi, Y. & Chiba K. (1998a). FTY720, a novel immunosuppressant, induces sequestration of circulating mature lymphocytes by acceleration of lymphocyte homing in rats. II. FTY720 prolongs skin allograft survival by preventing infiltration of T cells into the grafts, but not production of cytokines in vivo. *J. Immunol.*, Vol. 160, pp. 5493-5499.

Yanagawa, Y.; Masubuchi, Y. & Chiba K. (1998b). FTY720, a novel immunosuppressant, induces sequestration of circulating mature lymphocytes by acceleration of lymphocyte homing in rats. III. Increase in frequency of CD62L-positive T cells in Peyer's patches by FTY720-induced lymphocyte homing. *Immunology*, Vol. 95, pp. 591-594

Yanagawa, Y.; Hoshino, Y.; Kataoka, H.; Kawaguchi, T.; Ohtsuki, M.; Sugahara, K. & Chiba, K. (1999). FTY720, a novel immunosuppressant, prolongs rat skin allograft survival by decreasing T cell infiltration into grafts. *Transplant. Proc.*, Vol. 31, pp. 1227-1229

Yanagawa, Y.; Hoshino, Y. & Chiba, K. (2000). The significance of timing of FTY720 administration on the immunosuppressive effect to prolong rat skin allograft survival. *Int. J. Immunopharmacol.*, Vol. 22, pp. 597-602

Zamvil, S. S. & Steinman, L. (2003). Diverse targets for intervention during inflammatory and neurodegenerative phases of multiple sclerosis. *Neuron*, Vol. 38, pp. 685-688

6

Effects of Anxiolytic Drugs in Animal Models of Multiple Sclerosis

Silvia Novío, Manuel Freire-Garabal and María Jesús Núñez-Iglesias
Lennart Levi Stress and Neuroimmunology Laboratory,
University of Santiago de Compostela
Spain

1. Introduction

Multiple sclerosis (MS) is a chronic inflammatory demyelinating and neurodegenerative disease of the central nervous system (CNS), affecting more than 2 million people worldwide (Hirtz et al., 2007; McQualter & Bernard, 2007; Sospedra & Martin, 2005). Although it has been described for over two hundred years, it is not well characterized and no cure exists (Hirtz et al., 2007; McQualter & Bernard, 2007). For this reason, nowadays there is still considerable interest in the investigation of the pathogenesis of this disease, the improvement of diagnosis, the assessment of prognosis, and the discovery of new therapeutic agents.

The CNS cannot easily be sampled, so to gain ideas about neuroinflammatory diseases, animal models are developed. Experimental research has been performed in many species, including monkeys (Genain & Hauser, 2001), however most of the studies use rodents, fundamentally mice (Campbell et al., 2001; Chandler et al., 2002; Dowdell et al., 1999; Johnson et al., 2004, 2006; Meagher et al., 2007; Mi et al., 2004, 2006; Sieve et al., 2004, 2006; Steelman et al., 2009, 2010; Welsh et al., 2004; Whitacre et al., 1998; Young et al., 2008, 2010) and rats (Anane et al., 2003; Bukilica et al., 1991; Correa et al., 1998; Dimitrijević et al., 1994; Griffin et al., 1993; Kuroda et al., 1994; Laban et al., 1995a, 1995b; Le Page et al., 1994, 1996; Levine et al., 1962; Núñez-Iglesias et al., 2010; Owhashi et al., 1997; Pérez-Nievas et al., 2010; Teunis et al., 2002; Whitacre et al., 1998). In these studies, similar clinical phenotypes are achieved via different routes, so it is probable that some heterogeneity exists in the pathways leading to MS. In general, the standard experimental models of MS include: myelin mutant models, toxic demyelination models, viral models, and autoimmune models, being the virus-induced and immune-mediated models the most common ones for MS.

a. **Myelin mutant models.** Myelin mutants, such as the taiep rat and the *Shiverer* mouse (myelin basic protein (MBP) mutant), as well as gene knockout animals (e.g. myelin associated glycoprotein (MAG) knockout mouse) show axonal degeneration, altered neurotransmission, and in some instances clinical disease (Loers et al., 2004). Myelin mutant models have largely been used to study mechanisms of demyelination and remyelination. However, their relatively high cost has limited their widespread application (e.g. as preclinical drug screening tools).

b. **Toxic demyelination models.** Neurotoxicants such as lysolecithin, ethidium bromide or cuprizone are used to induce chemical lesions. In lysolecithin and ethidium bromide models, a focal lesion is induced by stereotactic injection of the compound into the rodent CNS. The toxic effect of lysolecithin is considered to be selective on myelin producing cells while ethidium bromide is toxic for all nucleolus containing cells (Woodruff & Franklin, 1999). The cuprizone model is widely used to study toxin induced demyelination. In this model, animals are fed with the copper chelator cuprizone (bis-cyclohexanone oxaldihydrazone) leading to demyelination, which is reversed after cessation of the toxin. This model is reliable and has the advantage of good reproducibility regarding the amount and site of demyelination (Matsushima & Morell, 2001).

c. **Viral models**. Several viruses, including Semliki Forest Virus and Theiler's Murine Encephalomyelitis Virus (TMEV), induce disease by neurotrophic infection of the CNS, specifically oligodendrocytes. Succinctly, virally-infected cells are attacked by T cells inducing important humoral responses, which finally lead to demyelination (Ercolini & Miller, 2006; Lavi & Constantinescu, 2005). The Picornavirus TMEV is a naturally occurring pathogen that was originally isolated from mice. In this species, strains of Theiler's virus (BeAn, DA, WW, Yale) cause a biphasic disease that includes an acute CNS inflammatory phase followed by a chronic neuroinflammatory/autoimmune demyelination phase with glial and microglial infection (Oleszak et al., 2004). The chronic phase of the disease has many similarities, both behaviorally and physiologically with progressive MS (Dal Canto et al., 1995; Lipton, 1975; Oleszak et al., 2004; Tsunoda & Fujinami, 1996), so Theiler's virus-induced demyelination (TVID) is commonly used as an excellent animal model of MS (Dal Canto et al., 1995; Oleszak et al., 2004; Tsunoda & Fujinami, 1996) for studying: the pathogenesis, the disease susceptibility factors, the mechanisms of viral persistence within the CNS, and the mechanisms of virus-induced autoimmune disease (Welsh et al., 2009).

d. **Autoimmune models**. Experimental autoimmune encephalomyelitis (EAE) has received the most attention as a model of MS. Clinical and histological features of MS can be actively or passively induced. Active EAE is accomplished through inoculation with spinal cord homogenate or with many different CNS proteins or peptides (such as myelin oligodendrocyte glycoprotein (MOG), myelin-associated oligodendrocyte basic protein (MOBP), oligodendrocyte-specific protein (OSP), proteolipid protein (PLP), and MBP) emulsified in adjuvant (e.g. complete Freund´s adjuvant (CFA), Pertussis toxin, alum, etc) (Tsunoda & Fujinami, 1996). Adjuvants potentiate immune reactions (Lavi & Constantinescu, 2005), ensure persistence of antigens at relevant sites (Lavi & Constantinescu, 2005), and influence stress response pathways inducing changes in levels of hormones such as ACTH (Selgas et al., 1997), so they can modulate the clinical course of EAE (Libbey & Fujinami, 2011). On the other hand, passive EAE is induced through adoptive transfer of myelin specific T cells into naïve animals (Tsunoda & Fujinami, 1996). Both models of EAE induction have been used extensively, with the active model most useful for studying the parameters involved in the initiation of EAE, and the passive model generally used in the study of the effector phase of EAE (Dittel et al., 1999). EAE is polygenic and the susceptibility and the clinical course (acute relapsing, chronic relapsing, relapsing-remitting, chronic progressive) can vary depending on the chosen EAE model and the strain/species of animal being investigated (Lavi & Constantinescu, 2005; Libbey & Fujinami, 2011; T. Owens, 2006).

Therefore, EAE is not a single model, but a number of models that have varying degrees of similarity to MS (Lavi & Constantinescu, 2005).

Some authors have doubts about the validity of experimental models of MS. However, at present it is accepted that although the preclinical research in MS is merely exploratory, it is also very necessary because it has contributed to elucidating key targets in the pathogenesis of MS. They have helped in the discovery of numerous cytokines and chemokines and the characterization of T helper cell subsets, thus playing a key role in understanding basic principles of immune function and autoimmunity (Gold et al., 2006). On the other hand, diagnostic, prognostic, and therapeutic aspects of MS have been cleared and resolved by means of experimental models (Pahan, 2010; Steinman & Zamvil, 2006). In this way, studies on EAE have culminated in three MS therapies (Steinman & Zamvil, 2006). For example glatiramer acetate, which was approved in 1996 for treatment of relapsing-remitting MS, currently is one of the most popular medications for treatment of relapsing-remitting MS, and more than 100,000 individuals with MS worldwide have received glatiramer acetate treatment (Sela, 2006). Besides, nowadays one of the exciting directions in the development of therapy for MS is consideration of various combinations of medications, and once again EAE models have demonstrated to be a valuable tool. They have shown potential synergies between drugs (statins and glatiramer), which show efficacy when used at doses that are suboptimal for these drugs when used alone (Greenwood et al., 2006; Stüve et al., 2006).

2. Stress and multiple sclerosis

The etiology of MS remains unknown, but studies have implicated both genetic and environmental factors (Noseworthy et al., 2000; Sospedra & Martin, 2005). The notion that psychological stress may be related to MS dates back to the time of Charcot, who suggested that the onset of MS is often preceded by grief or vexation, as well as by other socially undesirable circumstances (Charcot, 1877). Many studies since then have found that MS patients, as compared to healthy people or patients with other neurological disorders, report more stressful experiences prior to initial symptomatology. In the 1980s, two controlled studies were published on this issue. Their results showed that MS patients experienced remarkable life stress more frequently than the control subjects in the year (or six months) prior to MS onset (Grant et al., 1989; Warren et al., 1982). In addition to MS onset, relapses have also been found associated with stressful events (Ackerman et al., 2002; Brown et al., 2005, 2006; Franklin et al., 1988; Golan et al., 2008; Grant et al., 1989; Li et al., 2004; Mohr et al., 2004; Sibley, 1997). Franklin et al. (1988) in a longitudinal prospective study on 55 MS patients, with a clinical evaluation every 4 months for about 2 years, found that patients who reported significant negative or stressful life events were 3.7-times more likely to have an exacerbation than those free of such events. Sibley (1997) also found a significant association ($p<0.02$) between conjugal or job stress and MS relapses; in the same way that Mohr et al. (2004) in a systematic meta-analysis of 14 prospective studies, published from 1965 to 2003, found that there was a significantly increased risk of exacerbation associated with stressful life events (effect size of d=0.53; C.I.=0.40 to 0.65). In line with previously related studies, this relation has been further cleared by imaging techniques (magnetic resonance imaging) with the marker of acute focal brain inflammation, gadolinium (Goodin et al., 1999). In this way, Mohr et al. (2000) studied a group of 36 MS patients, finding that the occurrence of stressful life events was associated with a significantly increased risk of new gadolinium-enhancing (Gd+) brain lesions. Taken together, these findings and similar observations discovered in animal investigations (Campbell et al., 2001; Chandler et al., 2002; Johnson et al., 2004; Laban et al., 1995a; Meagher et al., 2007; Mi et al., 2004, 2006; Núñez-Iglesias et al., 2010; Pérez-Nievas et al., 2010; Sieve et al., 2004, 2006; Steelman et al., 2009, 2010; Teunis et al., 2002; Welsh et al., 2004; Young et al., 2008, 2010) confirm the necessity of applying preventive and tailored interventions, behavioral and pharmacological, in stressed patients with MS (Golan et al., 2008).

Despite all studies previously commented, some researchers have doubted about the association between the occurrence of stressful life events and the subsequent development of MS disease activity. Pratt (1951) and Gasperini et al. (1995) have not found significant differences between MS patients and control subjects, as far as their experienced stressful events were concerned; and even Nisipeanu and Korczyn (1993) have suggested that psychological stressors could have a "protective effect". Initially it was said that this discrepancy might be the result of a number of research design problems, including infrequent monitoring of patients, small patient samples, subjective reporting bias, type of statistical analysis used, lack of adequate controls, etc (Golan et al., 2008; Goodin, 2008; Martinelli, 2000). However, nowadays it is accepted that the relationship between MS and stressful life events is complex (Brown et al., 2005; Mohr et al., 2000). The type, the timing, and duration of the stressor as well as the animal strain and sex, and the chosen experimental model of MS (Mohr et al., 2004) are factors which determine the result:

a. **Type** (Table 1): Johnson et al. (2004) observed that if social stress is applied concurrently with Theiler's virus infection, disease severity is reduced compared to infected, non-stressed animals. In contrast, if restraint stress is applied concurrent with infection, the disease is again exacerbated (Campbell et al., 2001; Sieve et al., 2004). Likewise, Bukilica et al. (1991) indicated that whereas 19 daily sessions of inescapable tail-shock (80, 5 s, 1 mA) have no effect when administered prior to EAE induction, stressor exposure following EAE induction has a protective effect. Specifically, tail-shock reduces the incidence and duration of EAE, delays disease onset, and decreases the severity of clinical and histological symptoms.
b. **Timing and duration** (Table 1): Repeated moderate stressors suppress clinical signs when they are given before EAE induction, whereas acute severe stressors enhance the progression of disease after its induction (Heesen et al., 2007). Alternatively, acute stress applied prior to induction of EAE increases the severity of the disease (Teunis et al., 2002), and the contrary (i.e. a protective effect) is observed if the stressor is chronic (Levine & Saltzman, 1987; Levine et al., 1962; Whitacre et al., 1998).
c. **Animal strain** (Table 1): Certain inbred mouse strains, including SJL and DBA/2, are very susceptible to persistent CNS infection with TMEV and to the development of TVID, whereas other strains are intermediately susceptible (C3H, AKR, and CBA), and others are still able to clear the virus from the CNS, being resistant to the demyelinating phase of the disease (BALB/c and C57BL/6) (Sieve et al., 2004, 2006; Welsh et al., 1990). For example, Sieve et al. (2004, 2006) have observed important differences between CBA and SJL mice. Concretely: first, SJL mice show symptoms of the chronic phase of disease earlier (at 35 days pi) than CBA mice (at 150 days pi); second, SJL mice gradually develop late disease, whereas CBA mice have a sudden onset of severe symptoms; third, the incidence of the chronic phase is higher in SJL than in CBA mice (100% of the

Study	Model of MS	Stressor characteristics (type, timing, and duration)	Results	
			Acute stressor*	Chronic stressor**
Pérez-Nievas et al., 2010	DA rats, EAE (MOG/CFA)	Restraint stress started the same day of induction. Duration: 12d or 21d		Exacerbation (12d) Protective (21d)
Núñez-Iglesias et al., 2010	Lewis rats, EAE (MBP/CFA)	Noise stress started 5d prior to induction. Duration: 19d or 39d		Exacerbation (19d or 39d)
Young et al., 2010	SJL/JCrHsd mice, Theiler (BeAn strain)	Restraint stress started the day prior to infection. Duration: 28d		Exacerbation
Steelman et al., 2010	C57BL/6 mice, Theiler (BeAn strain)	Restraint stress started the day prior to infection. Duration: 7d or 4w		---
Steelman et al., 2009	SJL mice, Theiler (BeAn strain)	Restraint stress started the day prior to infection. Duration: 8d		Exacerbation
Young et al., 2008	CBA mice, Theiler (BeAn strain)	Restraint stress started the day prior to infection. Duration: 4w		Exacerbation
Meagher et al., 2007	Balb/cJ mice, Theiler (BeAn strain)	Social disruption stress started 1w before infection. Duration: 7d		Exacerbation
Mi et al., 2004, 2006	CBA mice, Theiler (BeAn strain)	Restraint stress started the day prior to infection. Duration: 2 or 7 d	Exacerbation	Exacerbation
Sieve et al., 2006	CBA mice, Theiler (BeAn strain)	Restraint stress started the day prior to infection. Duration: 4w		Exacerbation
Johnson et al., 2004	Balb/cJ mice, Theiler (BeAn strain)	Social disruption stress started: * 1w prior to infection or * the day of infection Duration: 7d		Exacerbation (stress applied prior to infection) Protective (stress applied concurrent with infection)
Sieve et al., 2004	SJL mice, Theiler (BeAn strain)	Restraint stress started the day prior to infection. Duration: 4w		Exacerbation
Welsh et al., 2004	CBA mice, Theiler (BeAn strain)	Restraint stress started the day prior to infection. Duration: 4w		Exacerbation
Anane et al., 2003	Lewis rats, EAE (MBP/CFA)	Physical stress (microwaves) started the day of induction. Duration: 21d		---
Chandler et al., 2002	SJL/J mice, EAE (PLP/CFA)	Restraint stress was performed on days 2 and 3 post-induction. Duration: 2d	Exacerbation	

Study	Model of MS	Stressor characteristics (type, timing, and duration)	Results	
			Acute stressor*	Chronic stressor**
Teunis et al., 2002	Wistar rats, EAE (MBP/CFA)	Neonatal maternal deprivation was performed aprox. 7w before induction. Duration: 24h	Exacerbation	
Campbell et al., 2001	CBA mice, Theiler (BeAn strain)	Restraint stress started the day prior to infection. Duration: 4w		Exacerbation
Dowdell et al., 1999	B10.PL mice, EAE (MBP/CFA)	Restraint stress started the day prior to induction. Duration: 21d		Protective
Whitacre et al., 1998	Lewis rats, EAE (MBP/CFA)	Restraint stress started 5 days prior to induction. Duration: 23d		Protective (9h of stress/d) Exacerbation (1 or 12h of stress/d)
Correa et al., 1998	Wistar rats, EAE (MBP/CFA)	Varied stress (swimming, predator odor, water deprivation, crowding, restraint, high-intensity sound, and cage inclination) was performed for the 14d before or after induction. Duration: 2w		Protective (stress before induction) Exacerbation (stress after induction)
Owhashi et al., 1997	Lewis rats, EAE (MBP/CFA)	Water bath (44ºC) was performed for the 10 or 13d before or after the induction. Duration: 10 or 13d	--- (stress before induction) Protective (stress after induction)	
Le Page et al., 1996	Lewis rats, adoptive EAE	Physical exercise was performed the 2d before or after the adoptive transfer of EAE. Duration: 2d	--- (stress before induction) Scantily protective (stress after induction)	
Laban et al., 1995a	DA rats, EAE (SCH/CFA)	Neonatal handling or gentling was performed 8w before induction. Duration: 4w	Exacerbation	
Laban et al., 1995b	DA rats, EAE (SCH/CFA)	Maternal deprivation was performed 8w before induction. Duration: 28d Early weaning was performed for 5-6w before induction. Duration: 1-2w		Protective (maternal deprivation Exacerbation (early weaning)

Study	Model of MS	Stressor characteristics (type, timing, and duration)	Results	
			Acute stressor*	Chronic stressor**
Le Page et al., 1994	Lewis rats, EAE (SCH/CFA)	Physical exercise was performed for the 10 days after induction. Duration: 10d		Protective
Dimitrijević et al., 1994	Lewis and DA rats, EAE (SCH/CFA)	Neonatal sound stress was performed 2 or 3w before induction. Duration: 1h	Exacerbation (Lewis) Protective (DA)	
Kuroda et al., 1994	Lewis rats, EAE (SCH/CFA)	Restraint stress was performed 1 or 8d after induction. Duration: 3d	--- (1d) Protective (8d)	
Griffin et al., 1993	Lewis rats, EAE (MBP/CFA)	Restraint stress started 5d before induction. Duration: 23d		Protective
Bukilica et al., 1991	DA rats, EAE (SCH/CFA)	Electric stress or sound stress was performed the 19d before or after induction. Duration: 19d		Protective (electric stress after induction and sound stress) --- (electric stress before induction)

Table 1. Animal studies (published from 1991 to 2010) on the effects of stress on disease manifestation. Studies are classified according to the type of stressor used: acute or chronic. *Acute stressor, stressor lasting less than 1 h and for less than 5 days; **chronic stressor, stressor lasting longer than 1 h and more than 5 days (although in most instances they were not presented all through the day). Abbreviations. CFA, complete Freund's adjuvant; d, day; DA, Dark August; EAE, experimental autoimmune encephalomyelitis; h, hour; MBP, myelin basic protein; MOG, myelin oligodendrocyte glycoprotein; PLP, proteolipid protein; SCH, spinal cord homogenate; w, week.

SJL mice develop severe symptoms of the chronic phase of the disease, versus to 70% (at most) of the CBA mice) (Friedmann & Lorch, 1985; Oleszak et al., 2004; Simas & Fazakerley, 1996). Susceptibility to TMEV persistence and TVID has been linked to genetic differences between strains of mice (Bureau et al., 1993; Monteyne et al., 1997; Oleszak et al., 2004; Rodriguez et al., 1990), which could explain the variability in their responsivity to stress and their different immunological background. In relation to EAE, it has also been shown that the susceptibility varies depending on the strain. So, whereas ABH and SJL mice develop relapsing EAE to disease induced by whole myelin, C57BL/6 mice are resistant (Lavi & Constantinescu, 2005).

d. **Sex**: A very discussed topic has been the sex impact in the disease process (Hill et al., 1998; Kappel et al., 1990; Lipton, 1975; Sieve et al., 2004, 2006). In some studies, female SJL mice are known to have greater susceptibility to disease as compared to males, a pattern that is similar to that found in human MS patients (Hill et al., 1998; Kappel et al., 1990; Sieve et al., 2004); on the contrary, other studies indicate that male mice develop more severe symptomatology of disease than females (Alley et al., 2003). It has been suggested that these apparently contradictory results may be due to different study designs and criteria used such as housing conditions or strain of Theiler's virus (Sieve et al., 2004). However, the sexual dimorphism of the immune system, the stress systems or the bidirectional communication between the reproductive system and the stress systems are reasons which may also explain, at least in part, this discrepancy (Gaillard & Spinedi, 1998; Whitacre et al., 1999). On the other hand, it is important to emphasize that the pattern of sex differences found can be complex. Sometimes, there are no sex differences in the early viral infection, existing on the contrary, greater behavioral signs in males than in females in later disease (Sieve et al., 2004).

e. **Experimental model of MS** (Table 1): Several authors have observed that stress exacerbates the early viral infection (Campbell et al., 2001; Sieve et al., 2004) and the later demyelinating disease (Sieve et al., 2004) in Theiler's virus infection. However, this does not coincide with studies using EAE, which show no effect of stress prior to disease induction, and a suppression during disease induction (Bukilica et al., 1991; Dowdell et al., 1999; Griffin et al., 1993; Levine & Saltzman, 1987; Levine et al., 1962). The differences in how stress affects EAE and Theiler's virus infection may lie in their immunological mechanisms of demyelination and neuronal destruction. However, the observed discrepancy between these two experimental models of MS can also be attributed to the fact that the stressor is applied during different phases in the immunological response of the disease process (Sieve et al., 2004).

3. Effects and mechanisms of action of benzodiazepines on models of MS

Benzodiazepines (alone or in association with other therapies) have long been used to relieve or resolve symptoms and signs associated with MS (Arroyo et al., 2011; Bush et al., 1996; D'Aleo et al., 2011; Hung & Huang, 2007; Meythaler et al., 1991; Rode et al., 2003; Solaro et al., 2010; Stork & Hoffman, 1994; Velez et al., 2003; Yerdelen et al., 2008). For example, Hung and Huang (2007) have observed that a combination of lorazepam and diazepam may be considered to release catatonic features in patients with MS, although the prescription of benzodiazepines associated with electroconvulsive therapy is another therapeutic option commonly used (Bush et al., 1996; Hung & Huang, 2007). Likewise, painful spasms, tremors or seizures (with or without associated anxiety symptoms) are treated with benzodiazepines such as clonazepam (Rode et al., 2003; Yerdelen et al., 2008), diazepam (D'Aleo et al., 2011; Meythaler et al., 1991; Rode et al., 2003) or tetrazepam (Rode et al., 2003); and even, Velez et al. (2003) have observed that patients with dramatic opisthotonic posturing and vermiform tongue fasciculations respond well to intravenous doses of lorazepam.

Benzodiazepines are used clinically as tranquilizers, muscle relaxants, anticonvulsants, anxiolytics, and sedative-hypnotics. These effects are mediated primarily via the central benzodiazepine receptors (CBR) located in the CNS (Heiss & Herholz, 2006); however, in addition to binding of $GABA_A$ receptors in the CNS, benzodiazepines bind to another site in peripheral tissues. This second type of recognition sites was mistakenly termed "peripheral benzodiazepine receptor" (PBR) for many years (Table 2). However, at present scientists prefer using the nomenclature: translocator protein (18 kDa) (TSPO) (Papadopoulos et al., 2006a). The TSPO is different from the CBR in terms of function, structure, expression, and pharmacological action (Gavish et al., 1999; Woods & Williams, 1996), so their study must be performed separately.

3.1 Effects mediated by the CBR

To date, only one study has been conducted to examine the influence of central benzodiazepine agonists on the development of animal models of MS (Núñez-Iglesias et al., 2010). Núñez-Iglesias et al. (2010) have observed that alprazolam decreases the clinical (paralysis, paraplegia, piloerection, etc) and histological (perivascular inflammatory infiltrate) manifestations of acute EAE in Lewis rats exposed to a chronic auditory stressor.

The molecular mechanisms mediating the clinical effects of central benzodiazepines in animal models of MS are unknown. However, it is thought that stress response mediators might play an important role in them.

	CBR	**TSPO**
Structure	Part of a macromolecular complex that also contains a γ-aminobutyric acid ($GABA_A$) receptor site and a chloride ion channel (Heiss & Herholz, 2006).	Part of a hetero-oligomeric complex comprised of the voltage-dependent anion channel and an adenine nucleotide carrier (McEnery et al., 1992; Papadopoulos et al., 2006a).
Subcellular localization	Plasma membrane of neurons (Heiss & Herholz, 2006).	Mitochondrial outer membrane (Gavish et al., 1999; Heiss & Herholz, 2006). Nonmitochondrial localization: plasma membrane (Gavish et al., 1999; Olson et al., 1988), nucleus, and perinuclear area (Gavish et al., 1999; Kuhlmann & Guilarte, 2000).
Localization	**Central**: medial occipital cortex, cerebellum, thalamus, striatum, pons (Heiss & Herholz, 2006).	**Peripheral**: kidney (Gavish et al., 1999), lung (Gavish et al., 1999), skeletal muscle (Gavish et al., 1999), liver (Gavish et al., 1999), heart (Gavish et al., 1999), uterus (Gavish et al., 1999), testis (Gavish et al., 1999), ovaries (Cosenza-Nashat et al., 2009; Gavish et al., 1999), haematogenous cells (Cosenza-Nashat et al., 2009; Olson et al., 1988; Ruff et al., 1985), and the steroid hormone-producing cells of the adrenal cortex (Gavish et al., 1999). **Central** (low concentrations): principally non-neuronal cells: ependymal lining of the ventricles, choroid plexus (Mattner et al., 2005), and glial cells (astrocytes and microglia) (Cosenza-Nashat et al., 2009; Mattner et al., 2005). Some studies also suggest that neurons may express TSPO (Jayakumar et al., 2002).

Table 2. Benzodiazepine binding sites. Main differences between CBR (central benzodiazepine receptor) and TSPO (translocator protein (18 kDa), also known as: peripheral-type benzodiazepine binding site, peripheral benzodiazepine receptor or mitochondrial benzodiazepine receptor).

3.1.1 Psychoneuroimmunoendocrinology and MS

Stress affects host defenses comprising neuronal, endocrine, and immune reactions. This complex network of bi-directional signals plays a vital role in determining the outcome of the stress response, since when the balance among these systems is altered, the risk of disease increases (Masood et al., 2003).

Figure 1 shows how stress impairs both natural and specific immune responses, which could influence morbidity associated with MS. Changes in the absolute number of lymphocytes, T-lymphocytes, T-helper, and T-suppressor cells have been reported (Freire-Garabal et al., 1991, 1997). Stress also interferes with several immune responses such as splenic cytotoxic activities, mediated by NK cells and cytotoxic T lymphocytes (Núñez et al., 2006), the activity of phagocytosis (Freire-Garabal et al., 1993a, 1993b), the delayed type hypersensitivity (DTH) response (Freire-Garabal et al., 1997; Núñez et al., 1998; Varela-Patiño et al., 1994), the blastogenic response of spleen lymphoid cells (Freire-Garabal et al., 1991, 1997), and T-dependent antibody responses (Fukui et al., 1997).

Research into the mechanisms by which the stressors are translated into impaired immune function and vulnerability to disease has focused primarily on two pathways: the hypothalamic-pituitary-adrenal (HPA) axis and the sympathetic branch of the autonomic nervous system (ANS) (Figure 1). Whereas increased sympathetic adrenal activity appears to play a major role in immune changes observed after acute stress, HPA axis-activity together with sympathetic mechanisms are mainly responsible for the inhibition of cellular and humoral immune responses following chronic stress exposure (Glaser & Kiecolt-Glaser, 2005). The importance of these systems is so high that when neuroendocrine hyper- or hypoactive responses of the HPA axis or the sympathetic nervous system (SNS) to stress occur, they function as risk factors of specific diseases, such as neurodegenerative diseases. Concretely, Gold et al. (2005) highlight the relevance of the functional status of the HPA axis in the control of EAE. During the experimentally induced disease in animals, the endogenous levels of glucocorticoids are elevated and the recovery from the disease is clearly dependent on this endocrine change (MacPhee et al., 1989). This endocrine response is immunologically mediated so it is mainly the result of the stimulation of the HPA axis by cytokines (such as IL-1) produced during the immune response that induces the autoimmune disease (Del Rey et al., 1998). In EAE models, the negative feedback system mediated via the glucocorticoid receptors seems to be disturbed (Gold et al., 2005), with the stressors favoring the perpetuation of this disregulation, as is shown by increased corticosterone levels in stressed rats relative to unstressed animals (Núñez-Iglesias et al., 2010). The importance of an increased HPA axis activity is supported by the observation that this phenomenon is related to the clinical disease course (Then Bergh et al., 1999).

The most basic literature regarding the HPA axis in pharmacology studies has been obtained in rats. More recently the mouse has been used due to the availability of genetically manipulated mice. The mouse is a model species of choice for genetic engineering because: a) its genes have an equivalent in humans; b) its genome is easy to modify by homologous recombination; c) it allows the creation of relevant animal models of human disease; d) numerous biological and biochemical functions of the mouse are similar to those of humans; e) it is easy to breed, less expensive to feed than rats and lives in smaller cages. These genetic models have allowed the determination of genes involved in anatomic and

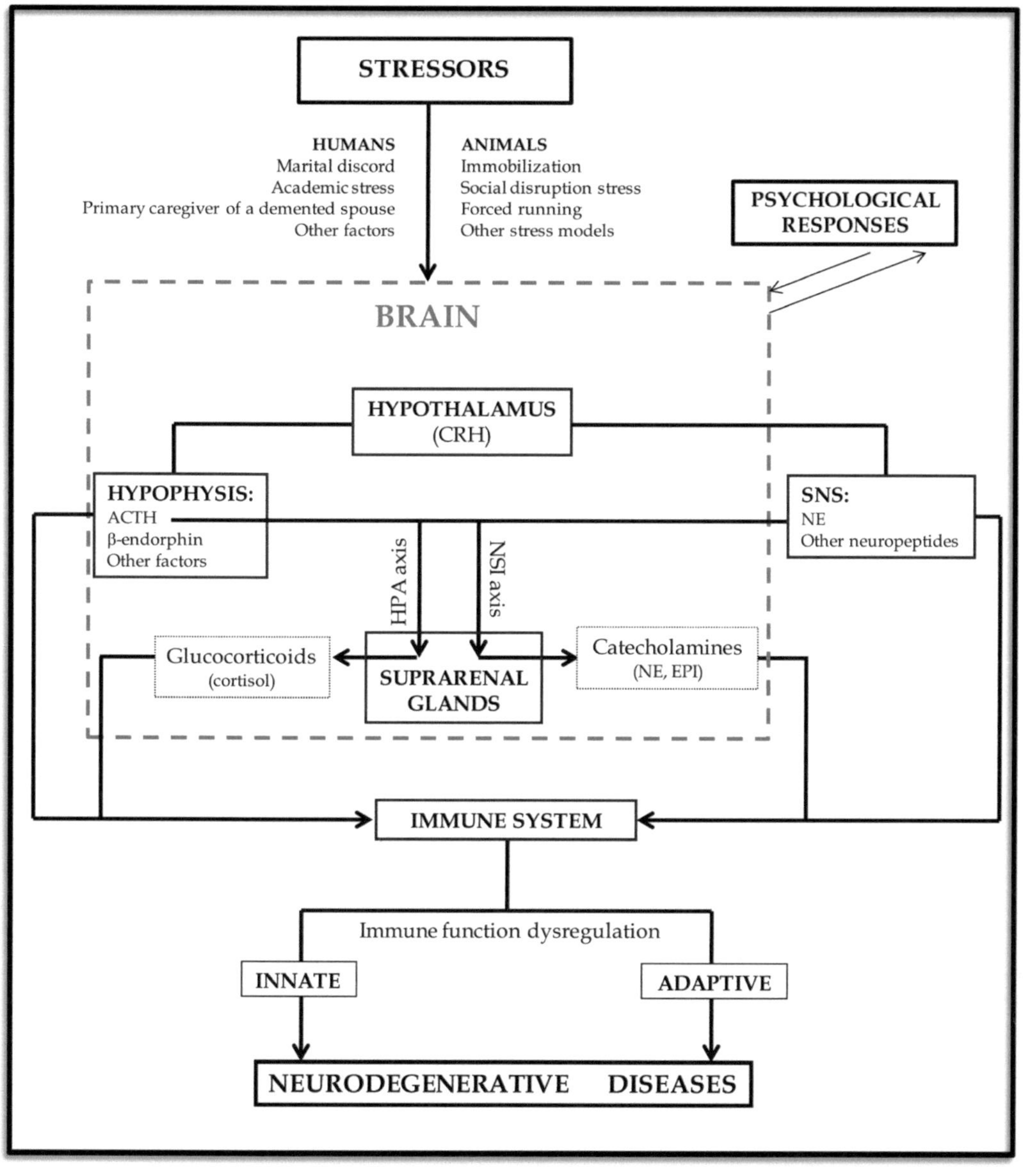

Fig. 1. Biological pathways mediating stress-induced modulation of immune function. The hypothalamic-pituitary-adrenal (HPA) axis and the neocortical-sympathetic-immune (NSI) axis are the main neural efferent pathways through which stress can affect the activity of the immune system. Stress-induced impairments in immunity can influence morbidity associated with neurodegenerative diseases. Abbreviations: ACTH, adrenocorticotropic hormone; CRH, corticotropin releasing hormone; EPI, epinephrine; NE, norepinephrine; SNS, sympathetic nervous system. Own production. Source: Friedman & Lawrence (2002) and Godbout & Glaser (2006).

functional alterations of brain circuits critical for stress regulation. Furthermore, they have contributed to understanding genetic vulnerability to anxiety and its pharmacological treatment (Gardier et al., 2009).

3.1.2 Benzodiazepines for control of stress associated with MS

Several mechanisms could explain the effects mediated by central benzodiazepines on attenuated manifestations of MS:

a. **Inhibitory influence on the activity of the HPA axis.**

GABA and benzodiazepines reduce levels of HPA axis hormones, including CRF (corticotropin releasing hormone) and ACTH (adrenocorticotropic hormone) and corticosterone (Arvat et al., 2002; Bizzi et al., 1984; M.J. Owens et al., 1989) acting on CBR. Central pharmacological effects related to CBR acting by facilitating inhibitory GABA neurotransmission in the CNS, may regulate the release of neuroendocrine hormones involved in the immune response to stress.

b. **Platelet activating factor (PAF) antagonist properties.**

Activation and control of the coagulation cascade, modulated by antigen-specific mediators of cellular immunity, appear to be of prime importance in the animal models of MS (Inoue et al., 1996). Susceptibility and resistance to EAE in rodents correlate with the induction of procoagulant and anticoagulant activities. Geczy et al. (1984) observed that anticoagulants produced by cells from nonsusceptible EAE rodents suppressed the common coagulation pathway by inhibiting trombin and factor Xa activities.

Central benzodiazepines such as alprazolam have PAF antagonist properties. It was found that in washed human platelets the alprazolam potently inhibits PAF-induced changes in shape, aggregation, and secretion, with the effects being specific for PAF activation (Kornecki et al., 1984). Likewise, Ng and Wong (1988) also showed that alprazolam can inhibit the [^{3}H]PAF binding to the human peripheral blood mononuclear leukocytes. In this context, it is interesting to point out that PAF plays a role in the activation of the HPA axis and glucocorticoid secretion and can serve as a mediator in the interactions of the immune system with the CNS. Concretely, PAF is an activator of the HPA axis in the rat. Its activation, which causes significant stimulation of hypothalamic CRH, pituitary ACTH, and adrenal corticosterone secretion, is inhibited by alprazolam. In addition, the PAF stimulates ACTH secretion by dispersed rat pituicytes, which is also inhibited by the alprazolam (Bernardini et al., 1989). The specific antagonism of PAF action by psychotropic drugs suggests that PAF or PAF-like phospholipids may play a role in neuronal function (Kornecki et al., 1987).

c. **Inhibitory activity on proinflammatory cytokines.**

Besides the mechanisms previously described, downstream effects of the alprazolam on immunological and inflammatory parameters important for EAE must be underscored. Secondarily recruited inflammatory cells account for the vast majority of infiltrating cells in MS lesions and they play a pivotal role in CNS tissue damage (Ransohoff, 1999). The detailed mechanisms by which inflammatory cells enter the CNS compartment are not completely understood. However, evidence suggests that cytokines are essential for this process (Karpus & Ransohoff, 1998). Enhanced expression of proinflammatory cytokines in the CNS, such as the monocyte chemoattractant protein 1 (MCP-1), has been demonstrated both in animal models of MS (Juedes et al., 2000) and in human case series (D'Aversa et al., 2002), and Karpus et al. (1997) have showed that the severity of manifestations is reduced by anti-MCP-1 antibodies. Additionally, mice that lack C-C chemokine receptor 2 (CCR2), the major receptor on monocytes for MCP-1, fail to develop the disease after active immunization (Fife et al., 2000) and are resistant to induction of it by the adoptive transfer of primed T cells from syngenic wild-type mice (Izikson et al., 2000). The effect of alprazolam on the expression levels of cytokines has been studied (Chang et al., 1992; Oda et al., 2002). Oda et al. (2002) have noted a potent inhibitory activity of this benzodiazepine on IL-1α-elicited MCP-1 production in T98G cells. Likewise, alprazolam inhibits the production of cytokines IL-1β and MCP-1 in LPS-stimulated mouse macrophage cells (Oda et al., 2002) and reduces the production of IL-2 by murine splenic T-cells (Chang et al., 1992). These findings suggest that alprazolam might prevent the infiltration of specific regions by an excess of proinflammatory cytokines. Since the excess production of proinflammatory cytokines exacerbates MS or EAE (Karpus & Ransohoff, 1998), the above-described action of alprazolam might explain the improvement of manifestations associated with EAE in non-human species or patients treated with this drug.

3.2 Effects mediated by the TSPO

Microglia play a significant role in the pathogenesis of MS (Venneti et al., 2006). They serve housekeeping functions and maintain homeostasis of local environments (Davalos et al., 2005; Nimmerjahn et al., 2005). In response to CNS insults, microglia change from a resting to an activated state to function as phagocytic macrophages (Chan et al., 2003; Fetler & Amigorena, 2005). This transition of microglia into an activated state includes a change in their morphology, migration towards the site of neuronal damage, proliferation until they quadruplicate in number (Davalos et al., 2005; Fetler & Amigorena, 2005), overexpression of cell markers (Agnello et al., 2000; Banati et al., 1997, 2000; Debruyne et al., 2003; Gavish et al., 1999; Kuhlmann & Guilarte, 2000; Versijpt et al., 2005; Vowinckel et al., 1997), and release of a widespread variety of substances or molecules (Chao et al., 1992, 1995a, 1995b; Colton et al., 1993; D'Aversa et al., 2002; Giulian et al., 1986, 1990; Heyes et al., 1996; McManus et al., 1998; Murphy et al., 1995; Righi et al., 1989). These findings demonstrate that microglia (together with perivascular macrophages -Guillemin & Brew, 2004-) represent a first line of the immune defense system of the brain (Davalos et al., 2005; Fetler & Amigorena, 2005; Nimmerjahn et al., 2005), and justify their description as a "sensor for pathological events in the CNS" (Kreutzberg, 1996). Parallel to this protective function, microglia can also contribute to aggravating the underlying neuronal damage via the synthesis and release of neurotoxins (Chao et al., 1992, 1995a; Colton et al., 1993; Giulian et al., 1990; Heyes et al., 1996), cytokines (Chao et al., 1995b; Giulian et al., 1986; Righi et al., 1989), and chemokines (D'Aversa et al., 2002; McManus et al., 1998; Murphy et al., 1995). Taking into account these results, it is concluded that microglia can exist in different states of activation depending on the microenvironment, with some states favoring the secretion of substances damaging neurons and other states favoring a protective phagocytic role (Morgan et al., 2005).

Microglia must maintain the balance between neurotoxicity and neuroprotection in injury, but the complex network of factors which governs their responses is only beginning to be deciphered (Biber et al., 2007; Glezer et al., 2007). Certainly, it would be interesting if some components of the network of microglial control could be manipulated for prognostic or therapeutic purposes of MS (Rock & Peterson, 2006). In this regard, TSPO plays a very important role. TSPOs are involved in the regulatory processes and metabolic functions of the tissue in which they are present. Outside the CNS: i) it is thought to aid in the transport of cholesterol from the outer to the inner mitochondrial membranes and thus be vital in steroid synthesis (Papadopoulos et al., 1997); ii) as a constituent of the mitochondrial permeability transition pore, TSPO is believed to regulate cell death (McEnery et al., 1992) and mitochondrial respiration (Hirsch et al., 1989); iii) evidence for an immunomodulatory role for this receptor includes the ability to: modulate chemotaxis and phagocytosis in peripheral monocytes and neutrophils (Marino et al., 2001; Ruff et al., 1985), induce cytokine expression and superoxide generation (Zavala et al., 1990), regulate macrophage functions (Pawlikowski, 1993), and stimulate formation of antibody-producing cells (Zavala & Lenfant, 1987), among others (Gavish et al., 1999); iv) TSPO is also thought to play a role in cell proliferation and differentiation (Camins et al., 1995), in protein and ion transport (Casellas et al., 2002; Gavish et al., 1999), and in bile acid synthesis (Lacapère & Papadopoulos, 2003; Woods & Williams, 1996). On the other hand, the functions of this receptor within the CNS are less known. It is suggested that it is involved in neurosteroid synthesis (Papadopoulos et al., 2006b), regulating mitochondrial function (Casellas et al., 2002), and modulating neuroinflammation in microglial cells (Wilms et al., 2003). The fact that TSPO knockout mice die at an early embryonic stage (Papadopoulos et al., 1997) strongly suggests that TSPO is involved in basic cell functions and is essential for embryonic development.

The main findings derived from the study of TSPO in MS patients or animal models of MS are detailed next:

a. **TSPO as *in vivo* marker of neuronal damage in MS.**

Reactive gliosis based on morphological examination is a microscopic finding in brain tissue sections and can only be obtained from invasive biopsy or postmortem autopsy. Therefore, the development and validation of an *in vivo* biomarker of glial damage is a major advance in the neurology field. In this way, the visualisation of the TSPO has received great importance in MS patients.

TSPO is expressed in the undamaged CNS at only a low level (Agnello et al., 2000; Banati et al., 2000; Gavish et al., 1999); however, its expression is dramatically increased (mainly on microglia and in minor importance on astrocytes) in inflammatory diseases such as MS (Banati et al., 2000; Debruyne et al., 2003; Versijpt et al., 2005; Vowinckel et al., 1997) and animal models of MS (Agnello et al., 2000; Banati et al., 2000; Gavish et al., 1999; Vowinckel et al., 1997). This up-regulation, which reflects an activation of resident microglia, can be visualized and measured using *in vitro* receptor autoradiography and binding assays as well as *in vivo* imaging techniques, such as PET (Maeda et al., 2004). So, in recent years a number of PET ligands with affinity to the TSPO have been developed and tested (e.g. Ro5-4864, PK11195, DAA1106, and vinpocetine) (Junck et al., 1989; Maeda et al., 2004). This has propitiated that nowadays TSPO cellular expression can be considered a reliable biomarker for neuroinflammation and gliosis with neuronal damage (Banati et al., 2000; Debruyne et al., 2003; Mattner et al., 2005; Versijpt et al., 2005; Vowinckel et al., 1997).

b. Neuroprotective function: anti-inflammatory and anti-apoptotic properties.

Recent evidence suggests that TSPO may play an important neuroprotective role in MS patients, both for its anti-inflammatory and its anti-apoptotic properties.

b.1 Anti-inflammatory properties.

Some investigations point to the possibility that the TSPO may participate actively in neuroinflammation and may thus itself be a target for therapeutic intervention. In this way, it has been demonstrated that TSPO ligands (Choi et al., 2002; Ryu et al., 2005) and some benzodiazepines (Wilms et al., 2003) possess anti-inflammatory properties. The PK11195 ligand inhibits increases in cyclooxygenase-2 levels in cultured human microglia (Choi et al., 2002), decreases expression of pro-inflammatory cytokines (IL-1β, IL-6, TNF-α) (Choi et al., 2002; Ryu et al., 2005) and reduces neuronal death in the quinolinic acid–injected rats (Ryu et al., 2005). Likewise, Wilms et al. (2003) have observed that midazolam, clonazepam, and diazepam interfere with the synthesis and release of proinflammatory (TNF-α) and neurotoxic (nitric oxide -NO-) molecules generated by activated microglia *in vitro*. The anti-inflammatory action associated to TSPO is not exclusive for microglial cells, it has also been shown on human blood cells (Bessler et al., 1992; Lenfant et al., 1986; Zavala et al., 1990). It is known that PK11195 and Ro5-4864 inhibit IL-3-like activity secretion in human peripheral blood mononuclear cells, and that IL-2, IL-1, TNF-α, and IL-6 production is inhibited by Ro5-4864 (Bessler et al., 1992; Lenfant et al., 1986). Likewise, treatment of mice with Ro5-4864 markedly reduces the capacity of macrophages to produce key mediators of inflammation such as reactive oxygen intermediates, IL-1, TNF, and IL-6 (Zavala et al., 1990). In particular, TNF is considered an important pharmacological target for the therapy of MS and drugs able to inhibit TNF-synthesis, such as the phosphodiesterase inhibitors, have been reported to ameliorate EAE (Sommer et al., 1995). Taken together, these findings are very promising, specially if we bear in mind that diazepam has been undoubtedly demonstrated to be neuroprotective in experimental models of other diseases (Schwartz-Bloom et al., 2000).

The true meaning of increased TSPO expression in microglia is unknown, however Wilms et al. (2003) have postulated that the presence of a high density of TSPO in human MS might be an adaptive response to neuronal damage with subsequent decreased release of neurotoxic microglial mediators. This hypothesis is supported by findings of Lacor et al. (1999) and Costa et al. (1994). They demonstrated that TSPO density is highly increased after peripheral nerve injury, with TSPO returning to normal levels when regeneration is complete or with TSPO remaining elevated in the absence of regeneration. A possible source of endogenous ligands of TSPO are astrocytes, which release substantial amounts of endozepines (Patte et al., 1999). These findings suggest that TSPO may be a trophic factor in recovery from brain injury.

A lot has been speculated about the mechanisms by which TSPO specific ligands confer protection. However, associations between TSPO activation and stimulation of neurosteroid synthesis have been noted (Lacapère & Papadopoulos, 2003; Le Goascogne et al., 2000). For example, Le Goascogne et al. (2000) have shown that TSPO activation in astrocytes promotes the synthesis of neurosteroids (Le Goascogne et al., 2000), which possess neurotrophic and neuroprotective activity (Le Goascogne et al., 2000) and are inhibitors of TNF production (Di Santo et al., 1996). A similar increase is obtained with anxiolytic benzodiazepines known to bind to both classes of benzodiazepine receptors (diazepam). On the contrary, ligands selective for the $GABA_A$ receptor (clonazepam) have no effect on steroid synthesis (Papadopoulos et al., 1992). On the other hand, Cascio et al. (2000) have shown a correlation among TSPO expression, steroid synthesis, myelination, and oligodendrocyte differentiation, thus reasserting the trophic function of TSPO in recovery from brain damage.

b.2 Anti-apoptotic properties.

The association of TSPO with the mitochondrial permeability transition pore suggests a role in the regulation of cell survival in microglia (McEnery et al., 1992). The participation of the TSPO in apoptotic processes has been demonstrated neither in MS patients nor in animal models of MS. However, studies that have induced overexpression of TSPO in cells different from those microglial ones suggest its implication in cell death regulation (Carayon et al., 1996; Everett & McFadden, 2001; Johnston et al., 2001; Rey et al., 2000; Stoebner et al., 2001). Interestingly, forced TSPO overexpression in myxoma poxvirus-infected macrophages blocks apoptosis (Everett & McFadden, 2001), in the same way that forced TSPO expression in neurons *in vivo* and Jurkat cells *in vitro* also protects these cells from apoptosis (Johnston et al., 2001; Stoebner et al., 2001). Likewise, it has been shown that TSPO upregulation in testicular Leydig cells (Rey et al., 2000) and in blood phagocytic cells (Carayon et al., 1996) preserves them from cytokine- and oxidant-induced cell death, respectively. TSPO expression in microglia may thus protect them from various toxins, thereby contributing to longer microglia life spans in the brain.

c. Neurotoxic effects.

A wealth of literature suggests that the TSPO overexpression , in addition to playing a protective role, can contribute to tissue destruction and disease progression (Block et al., 2007; Kreutzberg, 1996; Rothwell & Hopkins, 1995). When microglia enter an overactivated state, they synthesize and release a battery of potent neurotoxins (including free radicals (Block et al., 2007; Chao et al., 1995a), NO (Chao et al., 1992), proteinases (Colton et al., 1993), eicosanoids (Heyes et al., 1996), and excitotoxins (Giulian et al., 1990), cytokines (IL-1 (Giulian et al., 1986), IL-6 (Righi et al., 1989), and TNFα (Chao et al., 1995b)), and chemokines (such as MIP-1α (Murphy et al., 1995), MIP-1β (McManus et al., 1998), and MCP-1 (D'Aversa et al., 2002)) that cause neurotoxicity, influencing the viability and function of neurons and exacerbating neuronal injury. Two major possible neurotoxic secretion products of microglial cells are NO and TNF-α (Wilms et al., 2003). NO is neurotoxic due to inhibition of complex 1 and 2 of the respiratory chain. Moreover, it reacts with superoxide anion to generate peroxynitrite, a highly reactive molecule capable of oxidizing proteins, lipids, and DNA. The cytokine TNF-α is an important factor in the regulation of neuronal apoptotic cell death, which is expressed by astrocytes and microglial cells in brain lesions of MS patients (Wilms et al., 2003).

The inhibition of microglial activation by a pharmacological approach, using non-steroidal anti-inflammatory drugs or minocycline, has been hypothesized to reduce neuronal damage in animal models of neurodegenerative diseases (Du et al., 2001). Furthermore, activation of microglia also inhibits neurogenesis in the rat hippocampus, and hippocampal regeneration is restored by blocking microglial activation with either indomethacin (Monje et al., 2003) or minocycline (Ekdahl et al., 2003). These studies suggest that activation of microglia could perpetrate neurodegeneration through several mechanisms.

4. Conclusion

Animal models of MS are a very beneficial tool, which have led to a better understanding of MS. New clues to the pathogenesis of MS and new potential markers for the diagnosis and prognosis of MS have been gained from research in animal models. Likewise, they have helped in the development of therapeutic approaches that are currently being used.

The susceptibility to MS is modulated by interactions among many factors. In this context, it has been hypothesized that disease onset, progression, and relapses in MS are associated with stressful life events, and this alleged relation has been confirmed by sophisticated medical techniques. However, it is necessary to bear in mind that stressor characteristics are key factors in determining the effects of stress on MS symptom development.

Drugs known to affect the immune system have become the primary focus for managing MS. However, the most recent findings suggest that benzodiazepines might be an add-on option for MS treatment because they can modify the stress-induced manifestations of EAE by interacting with CBRs. Concretely, it has been demonstrated that alprazolam reduces the latent period and inflammatory lesions of the SNC and delays the onset of the disease. Several mechanisms have been hypothesized to explain the effects of this type of drugs, which influence hormonal, immune, endocrine, and/or inflammatory parameters associated with the HPA axis and the sympathetic branch of the ANS.

Recent evidence suggests that TSPOs might play a dual role in MS patients and perform neuroprotective and neurotoxic functions. On the other hand, because TSPO is dramatically up-regulated in MS, TSPO cellular expression is considered a reliable marker for diagnosis of the disease progression and of the therapeutic response.

5. References

Ackerman, K.D., Heyman, R., Rabin, B.S., Anderson, B.P., Houck, P.R., Frank, E., & Baum, A. (2002). Stressful life events precede exacerbations of multiple sclerosis. *Psychosomatic Medicine,* Vol.64, No.6, pp. 916-920

Agnello, D., Carvelli, L., Muzio, V., Villa, P., Bottazzi, B., Polentarutti, N., Mennini, T., Mantovani, A., & Ghezzi, P. (2000). Increased peripheral benzodiazepine binding sites and pentraxin 3 expression in the spinal cord during EAE: relation to inflammatory cytokines and modulation by dexamethasone and rolipram. *Journal of Neuroimmunology,* Vol.109, No.2, pp. 105-111

Alley, J., Khasabov, S., Simone, D., Beitz, A., Rodriguez, M., & Njenga, M.K. (2003). More severe neurologic deficits in SJL/J male than female mice following Theiler's virus-induced CNS demyelination. *Experimental Neurology,* Vol.180, No.1, pp. 14-24

Anane, R., Geffard, M., Taxile, M., Bodet, D., Billaudel, B., Dulou, P.E., & Veyret, B. (2003). Effects of GSM-900 microwaves on the experimental allergic encephalomyelitis (EAE) rat model of multiple sclerosis. *Bioelectromagnetics,* Vol.24, No.3, pp. 211-213

Arroyo, R., Vila, C., & Clissold, S. (2011). Retrospective observational study of the management of multiple sclerosis patients with resistant spasticity in Spain: the '5E' study. *Expert Review of Pharmacoeconomics & Outcomes Research,* Vol.11, No.2, pp. 205-213

Arvat, E., Giordano, R., Grottoli, S., & Ghigo, E. (2002). Benzodiazepines and anterior pituitary function. *Journal of Endocrinological Investigation,* Vol.25, No.8, pp. 735-747

Banati, R.B., Myers, R., & Kreutzberg, G.W. (1997). PK ('peripheral benzodiazepine')--binding sites in the CNS indicate early and discrete brain lesions: microautoradiographic detection of [3H]PK11195 binding to activated microglia. *Journal of Neurocytology,*Vol.26, No.2, pp. 77-82

Banati, R.B., Goerres, G.W., Myers, R., Gunn, R.N., Turkheimer, F.E., Kreutzberg, G.W., Brooks, D.J., Jones, T., & Duncan, J.S. (1999). [11C](R)-PK11195 positron emission tomography imaging of activated microglia in vivo in Rasmussen's encephalitis. *Neurology,* Vol.53, No.9, pp. 2199-2203

Banati, R.B., Newcombe, J., Gunn, R.N., Cagnin, A., Turkheimer, F., Heppner, F., Price, G., Wegner, F., Giovannoni, G., Miller, D.H., Perkin, G.D., Smith, T., Hewson, A.K., Bydder, G., Kreutzberg, G.W., Jones, T., Cuzner, M.L., & Myers, R. (2000). The peripheral benzodiazepine binding site in the brain in multiple sclerosis: quantitative in vivo imaging of microglia as a measure of disease activity. *Brain,* Vo.123, No.11, pp. 2321-2337

Bernardini, R., Calogero, A.E., Ehrlich, Y.H., Brucke, T., Chrousos, G.P., & Gold, P.W. (1989). The alkyl-ether phospholipid platelet-activating factor is a stimulator of the hypothalamic-pituitary-adrenal axis in the rat. *Endocrinology,* Vol.125, No.2, pp. 1067-1073

Bessler, H., Weizman, R., Gavish, M., Notti, I., & Djaldetti, M. (1992). Immunomodulatory effect of peripheral benzodiazepine receptor ligands on human mononuclear cells. *Journal of Neuroimmunology,* Vol.38, No.1-2, pp. 19-25

Biber, K., Neumann, H., Inoue, K., & Boddeke, H.W. (2007). Neuronal 'On' and 'Off' signals control microglia. *Trends in Neurosciences,* Vol.30, No.11, pp. 596-602

Bizzi, A., Ricci, M.R., Veneroni, E., Amato, M., & Garattini, S. (1984). Benzodiazepine receptor antagonists reverse the effect of diazepam on plasma corticosterone in stressed rats. *The Journal of Pharmacy and Pharmacology,* Vol.36, No.2, pp. 134-135

Block, M.L., Zecca, L., & Hong, J.S. (2007). Microglia-mediated neurotoxicity: uncovering the molecular mechanisms. *Nature Reviews Neuroscience,* Vol.8, No.1, pp. 57-69

Brown, R.F., Tennant, C.C., Dunn, S.M., & Pollard, J.D. (2005). A review of stress–relapse interactions in multiple sclerosis: important features and stress-mediating and -moderating variables. *Multiple Sclerosis,* Vol.11, No.4, pp. 477–484

Brown, R.F., Tennant, C.C., Sharrock, M., Hodgkinson, S., Dunn, S.M., & Pollard, J.D. (2006). Relationship between stress and relapse in multiple sclerosis: Part I. Important features. *Multiple Sclerosis,* Vol.12, No.4, pp. 453-464

Bukilica, M., Djordjević, S., Marić, I., Dimitrijević, M., Marković, B.M., & Janković, B.D. (1991). Stress-induced suppression of experimental allergic encephalomyelitis in the rat. *The International Journal of Neuroscience,* Vol.59, No.1-3, pp. 167-175

Bureau, J.F., Montagutelli, X., Bihl, F., Lefebvre, S., Guénet , J.L., & Brahic, M. (1993). Mapping loci influencing the persistence of Theiler's virus in the murine central nervous system. *Nature Genetics,* Vol.5, No.1, pp. 87-91

Bush, G., Fink, M., Petrides, G., Dowling, F., & Francis, A. (1996). Catatonia. II. Treatment with lorazepam and electroconvulsive therapy. *Acta Psychiatrica Scandinavica,* Vol.93, No.2, pp. 137-143

Camins, A., Diez-Fernandez, C., Pujadas, E., Camarasa, J., & Escubedo, E. (1995). A new aspect of the antiproliferative action of peripheral-type benzodiazepine receptor ligands. *European Journal of Pharmacology,* Vol.272, No.2-3, pp. 289-292

Campbell, T., Meagher, M.W., Sieve, A., Scott, B., Storts, R., Welsh, T.H., & Welsh, C.J. (2001). The effects of restraint stress on the neuropathogenesis of Theiler's virus infection: I. Acute disease. *Brain, Behavior, and Immunity,* Vol.15, No.3, pp. 235-254

Carayon, P., Portier, M., Dussossoy, D., Bord, A., Petitprêtre, G., Canat, X., Le Fur, G., & Casellas, P. (1996). Involvement of peripheral benzodiazepine receptors in the protection of hematopoietic cells against oxygen radical damage. *Blood,* Vol.87, No.8, pp. 3170-3178

Cascio, C., Brown, R.C., Liu, Y., Han, Z., Hales, D.B., & Papadopoulos, V. (2000). Pathways of dehydroepiandrosterone formation in rat brain glia. *The Journal of Steroid Biochemistry and Molecular Biology,* Vol.75, No.2-3, pp. 177-186

Casellas, P., Galiegue, S., & Basile, A.S. (2002). Peripheral benzodiazepine receptors and mitochondrial function. *Neurochemistry International,* Vol.40, No.6, pp. 475-486

Chan, A., Seguin, R., Magnus, T., Papadimitriou, C., Toyka, K.V., Antel, J.P., & Gold, R. (2003). Phagocytosis of apoptotic inflammatory cells by microglia and its therapeutic implications: termination of CNS autoimmune inflammation and modulation by interferon-beta. *Glia,* Vol.43, No.3, pp. 231-242

Chandler, N., Jacobson, S., Esposito, P., Connolly, R., & Theoharides, T.C. (2002). Acute stress shortens the time to onset of experimental allergic encephalomyelitis in SJL/J mice. *Brain, Behavior, and Immunity,* Vol.16, No.6, pp. 757-763

Chang, M.P., Castle, S.C., & Norman, D.C. (1992). Mechanism of immunosuppressive effect of alprazolam: alprazolam suppresses T-cell proliferation by selectively inhibiting the production of IL2 but not acquisition of IL2 receptor. *International Journal of Immunopharmacology,* Vol.14, No.2, pp. 227-237

Chao, C.C., Hu, S., Molitor, T.W., Shaskan, E.G., & Peterson, P.K. (1992). Activated microglia mediate neuronal cell injury via a nitric oxide mechanism. *Journal of Immunology,* Vol.149, No.8, pp. 2736-2741

Chao, C.C., Hu, S., & Peterson, P.K. (1995a). Modulation of human microglial cell superoxide production by cytokines. *Journal of Leukocyte Biology,* Vol.58, No.1, pp. 65-70

Chao, C.C., Hu, S., Sheng, W.S., & Peterson, P.K. (1995b). Tumor necrosis factor-alpha production by human fetal microglial cells: regulation by other cytokines. *Developmental Neuroscience,* Vol.17, No.2, pp. 97-105

Charcot, J.M. (1877). *Lectures on diseases of the nervous system,* New Sydenham Society, London

Choi, H.B., Khoo, C., Ryu, J.K., van Breemen, E., Kim, S.U., & McLarnon, J.G. (2002). Inhibition of lipopolysaccharide-induced cyclooxygenase-2, tumor necrosis factor-alpha and [Ca2+]i responses in human microglia by the peripheral benzodiazepine receptor ligand PK11195. *Journal of Neurochemistry,* Vol.83, No.3, pp. 546-555

Colton, C.A., Keri, J.E., Chen, W.T., & Monsky, W.L. (1993). Protease production by cultured microglia: substrate gel analysis and immobilized matrix degradation. *Journal of Neuroscience Research,* Vol.35, No.3, pp. 297-304

Conway, E.L., Gundlach, A.L., & Craven, J.A. (1998). Temporal changes in glial fibrillary acidic protein messenger RNA and [3H]PK11195 binding in relation to imidazoline-I2-receptor and alpha 2-adrenoceptor binding in the hippocampus following transient global forebrain ischaemia in the rat. *Neuroscience,* Vol.82, No.3, pp. 805-817

Correa, S.G., Rodriguez-Galán, M.C., Rivero, V.E., & Riera, C.M. (1998). Chronic varied stress modulates experimental autoimmune encephalomyelitis in Wistar rats. *Brain, Behavior, and Immunity,*Vol.12, No.2, pp. 134-148

Cosenza-Nashat, M., Zhao, M.L., Suh, H.S., Morgan, J., Natividad, R., Morgello, S., & Lee, S.C. (2009). Expression of the translocator protein of 18 kDa by microglia, macrophages and astrocytes based on immunohistochemical localization in abnormal human brain. *Neuropathology and Applied Neurobiology,* Vol.35, No.3, pp. 306-328

Costa, E., Auta, J., Guidotti, A., Korneyev, A., & Romeo, E. The pharmacology of neurosteroidogenesis. *The Journal of Steroid Biochemistry and Molecular Biology,* Vol.49, No.4-6, pp. 385-389

Dal Canto, M.C., Melvold, R.W., Kim, B.S., & Miller, S.D. (1995). Two models of multiple sclerosis: experimental allergic encephalomyelitis (EAE) and Theiler's murine encephalomyelitis virus (TMEV) infection. A pathological and immunological comparison. *Microscopy Research and Technique,* Vol.32, No.3, pp. 215-229

D'Aleo, G., Rifici, C., Kofler, M., Sessa, E., Saltuari, L., & Bramanti, P. (2011). Seizure after intrathecal baclofen bolus in a multiple sclerosis patient treated.with oxcarbazepine. *Neurological Sciences,* Vol.32, No.2, pp. 293-295

Davalos, D., Grutzendler, J., Yang, G., Kim, J.V., Zuo, Y., Jung, S., Littman, D.R., Dustin, M.L., & Gan, W.B. (2005). ATP mediates rapid microglial response to local brain injury in vivo. *Nature Neuroscience,* Vol.8, No.6, pp. 752-758

D'Aversa, T.G., Weidenheim, K.M., & Berman, J.W. (2002). CD40-CD40L interactions induce chemokine expression by human microglia: implications for human immunodeficiency virus encephalitis and multiple sclerosis. *The American Journal of Pathology,* Vol.160, No.2, pp. 559-567

Debruyne, J.C., Versijpt, J., Van Laere, K.J., De Vos, F., Keppens, J., Strijckmans, K., Achten, E., Slegers, G., Dierckx, R.A., Korf, J., & De Reuck, J.L. (2003). PET visualization of microglia in multiple sclerosis patients using [11C]PK11195. *European Journal of Neurology,* Vol.10, No.3, pp. 257-264

Del Rey, A., Klusman, I., & Besedovsky, H.O. (1998). Cytokines mediate protective stimulation of glucocorticoid output during autoimmunity: involvement of IL-1. *The American Journal of Physiology,* Vol.275, No.4, pp. R1146-1151

Di Santo, E., Sironi, M., Mennini, T., Zinetti, M., Savoldi, G., Di Lorenzo, D., & Ghezzi, P. (1996). A glucocorticoid receptor-independent mechanism for neurosteroid inhibition of tumor necrosis factor production. *European Journal of Pharmacology,* Vol.299, No.1-3, pp. 179-186

Dittel, B.N., Visintin, I., Merchant, R.M., & Janeway, C.A. Jr. (1999). Presentation of the self antigen myelin basic protein by dendritic cells leads to experimental autoimmune encephalomyelitis. *Journal of Immunology,* Vol.163, No.1, pp. 32-39

Dimitrijević, M., Laban, O., von Hoersten, S., Marković, B.M., & Janković, B.D. (1994). Neonatal sound stress and development of experimental allergic encephalomyelitis in Lewis and DA rats. *The International Journal of Neuroscience,* Vol.78, No.1-2, pp. 135-143

Dowdell, K.C., Gienapp, I.E., Stuckman, S., Wardrop, R.M., & Whitacre, C.C. (1999). Neuroendocrine modulation of chronic relapsing experimental autoimmune encephalomyelitis: a critical role for the hypothalamic-pituitary-adrenal axis. *Journal of Neuroimmunology,* Vol.100, No.1-2, pp. 243-251

Du, Y., Ma, Z., Lin, S., Dodel, R.C., Gao, F., Bales, K.R., Triarhou, L.C., Chernet, E., Perry, K.W., Nelson, D.L., Luecke, S., Phebus, L.A., Bymaster, F.P., & Paul, S.M. (2001). Minocycline prevents nigrostriatal dopaminergic neurodegeneration in the MPTP model of Parkinson's disease. *Proceedings of the Society for Experimental Biology and Medicine,* Vol.98, No.25, pp. 14669-14674

Ekdahl, C.T., Claasen, J.H., Bonde, S., Kokaia, Z., & Lindvall, O. (2003). Inflammation is detrimental for neurogenesis in adult brain. *Proceedings of the Society for Experimental Biology and Medicine,* Vol.100, No.23, pp. 13632-13637

Ercolini, A.M., & Miller, S.D. (2006). Mechanisms of immunopathology in murine models of central nervous system demyelinating disease. *Journal of Immunology,* Vol.176, No.6, pp. 3293-3298

Everett, H., & McFadden, G. (2001). Viruses and apoptosis: meddling with mitochondria. *Virology,* Vol.288, No.1, pp. 1-7

Fetler, L., & Amigorena, S. (2005). Brain Under Surveillance: The Microglia Patrol. *Science,* Vol.309, No.5733, pp. 392-393

Fife, B.T., Huffnagle, G.B., Kuziel, W.A., & Karpus, W.J. (2000). CC chemokine receptor 2 is critical for induction of experimental autoimmune encephalomyelitis. *The Journal of Experimental Medicine,* Vol.192, No.6, pp. 899-905

Franklin, G.M., Nelson, L.M., Heaton, R.K., Burks, J.S., & Thompson, D.S. (1988). Stress and its relationship to acute exacerbations in multiple sclerosis (MS). *The Journal of Neurological Rehabilitation,* Vol.2, No.1, pp. 7-11

Freire-Garabal, M., Belmonte, A., & Suárez-Quintanilla, J. (1991). Effects of buspirone on the immunosuppressive response to stress in mice. *Archives Internationales de Pharmacodynamie et de Thérapie,* Vol.314, No.1-2, pp. 160-168

Freire-Garabal, M., Núñez, M.J., Balboa, J.L., González-Bahillo, J., & Belmonte, A. (1993a). Effects of midazolam on the activity of phagocytosis in mice submitted to surgical stress. *Pharmacology, Biochemistry and Behavior,* Vol.46, No.3, pp. 605-608

Freire-Garabal, M., Núñez, M.J., Fernández-Rial, J.C., Couceiro, J., García-Vallejo, L., & Rey-Méndez, M. (1993b). Phagocytic activity in stressed mice: effects of alprazolam. *Research in Immunology,* Vol.144, No.5, pp. 311-316

Freire-Garabal, M., Núñez, M.J., Losada, C., Pereiro, D., Riveiro, M.P., González-Patiño, E., Mayán, J.M., & Rey-Mendez, M. (1997). Effects of fluoxetine on the immunosuppressive response to stress in mice. *Life Sciences,* Vol.60, No.26, pp. PL403-413

Friedman, E.M., & Lawrence, D.A. (2002). Environmental stress mediates changes in neuroimmunological interactions. *Toxicological Sciences,* Vol.67, No.1, pp. 4-10

Friedmann, A., & Lorch, Y. (1985). Theiler's virus infection: a model for multiple sclerosis. *Progress in Medical Virology,* Vol.31, pp. 43-83

Fukui, Y., Sudo, N., Yu, X.N., Nukina, H., Sogawa, H., & Kubo, C. (1997). The restraint stress-induced reduction in lymphocyte cell number in lymphoid organs correlates with the suppression of in vivo antibody production. *Journal of Neuroimmunology,* Vol.79, No.2, pp. 211-217

Gaillard, R.C., & Spinedi, E. (1998). Sex- and stress-steroids interactions and the immune system: evidence for a neuroendocrine-immunological sexual dimorphism. *Domestic Animal Endocrinology,* Vol.15, No.5, pp. 345-352

Gardier, A.M., Guiard, B.P., Guilloux, J.P., Repérant, C., Coudoré, F., & David, D.J. (2009). Interest of using genetically manipulated mice as models of depression to evaluate antidepressant drugs activity: a review. *Fundamental & Clinical Pharmacology,* Vol.23, No.1, pp. 23-42

Gasperini, C., Grasso, M.G., Fiorelli, M., Millefiorini, E., Morino, S., Anzini, A., Colleluori, A., Salvetti, M., Buttinelli, C., & Pozzilli, C. (1995). A controlled study of potential risk factors preceding exacerbation in multiple sclerosis. *Journal of Neurology, Neurosurgery, and Psychiatry,* Vol.59, No.3, pp. 303-305

Gavish, M., Bachman, I., Shoukrun, R., Katz, Y., Veenman, L., Weisinger, G., & Weizman, A. (1999). Enigma of the peripheral benzodiazepine receptor. *Pharmacologica Reviews,* Vol.51, No.4, pp. 629-650

Geczy, C.L., Roberts, I.M., Meyer, P., & Bernard, C.C. (1984). Susceptibility and resistance to experimental autoimmune encephalomyelitis and neuritis in the guinea pig correlate with the induction of procoagulant and anticoagulant activities. *Journal of Immunology,* Vol.133, No.6, pp. 3026-3036

Genain, C.P., & Hauser, S.L. (2001). Experimental allergic encephalomyelitis in the New World monkey Callithrix jacchus. *Immunological Reviews,* Vol.183, No.1, pp. 159-172

Giulian, D., Baker, T.J., Shih, L.C., & Lachman, L.B. (1986). Interleukin 1 of the central nervous system is produced by ameboid microglia. *The Journal of Experimental Medicine,* Vol.164, No.2, pp. 594-604

Giulian, D., Vaca, K., & Noonan, C.A. (1990). Secretion of neurotoxins by mononuclear phagocytes infected with HIV-1. *Science,* Vol.250, No.4987, pp. 1593-1596

Glaser, R., & Kiecolt-Glaser, J.K. (2005). Stress-induced immune dysfunction: implications for health. *Nature Reviews. Immunology,* Vol.5, No.3, pp. 243-251

Glezer, I., Simard, A.R., & Rivest, S. (2007). Neuroprotective role of the innate immune system by microglia. *Neuroscience,* Vol.147, No.4, pp. 867-883

Godbout, J.P., & Glaser, R. (2006). Stress-induced immune dysregulation: implications for wound healing, infectious disease and cancer. *Journal of Neuroimmune Pharmacology,* Vol.1, No.4, pp. 421-427

Golan, D., Somer, E., Dishon, S., Cuzin-Disegni, L., & Miller, A. (2008). Impact of exposure to war stress on exacerbations of multiple sclerosis. *Annals of Neurology,* Vol.64, No.2, pp. 143-148

Gold, R., Linington, C., & Lassmann, H. (2006). Understanding pathogenesis and therapy of multiple sclerosis via animal models: 70 years of merits and culprits in experimental autoimmune encephalomyelitis research. *Brain,* Vol.129, No.8, pp. 1953-1971

Gold, S.M., Mohr, D.C., Huitinga, I., Flachenecker, P., Sternberg, E.M., & Heesen, C. (2005). The role of stress-response systems for the pathogenesis and progression of MS. *Trends in Immunology,* Vol.26, No.12, pp. 644-652

Goodin, D.S. (2008). The impact of war-stress on MS exacerbations. *Annals of Neurology,* Vol.64, No.2, pp. 114-115

Goodin, D.S., Ebers, G.C., Johnson, K.P., Rodriguez, M., Sibley, W.A., & Wolinsky, J.S. (1999). The relationship of MS to physical trauma and psychological stress: report of the Therapeutics and Technology Assessment Subcommittee of the American Academy of Neurology. *Neurology,* Vol.52, No.9, pp. 1737-1745

Grant, I., Brown, G.W., Harris, T., McDonald, W.I., Patterson, T., & Trimble, M.R. (1989). Severely threatening events and marked life difficulties preceding onset or exacerbation of multiple sclerosis. *Journal of Neurology, Neurosurgery and Psychiatry,* Vol.52, No.1, pp. 8-13

Greenwood, J., Steinman, L., & Zamvil, S.S. (2006). Statin therapy and autoimmune disease: from protein prenylation to immunomodulation. *Nature Reviews. Immunology,* Vol.6, No.5, pp. 358-370

Griffin, A.C., Lo, W.D., Wolny, A.C., & Whitacre, C.C. (1993). Suppression of experimental autoimmune encephalomyelitis by restraint stress: sex differences. *Journal of Neuroimmunology,* Vol.44, No.1, pp. 103-116

Guillemin, G.J., & Brew, B.J. (2004). Microglia, macrophages, perivascular macrophages, and pericytes: a review of function and identification. *Journal of Leukocyte Biology,* Vol.75, No.3, pp. 388-397

Heesen, C., Gold, S.M., Huitinga, I., & Reul, J.M. (2007). Stress and hypothalamic-pituitary-adrenal axis function in experimental autoimmune encephalomyelitis and multiple sclerosis - a review. *Psychoneuroendocrinology,* Vol.32, No.6, pp. 604-618

Heiss, W.D., & Herholz, K. (2006). Brain receptor imaging. *Journal of Nuclear Medicine,* Vol.47, No.2, pp. 302-312

Heyes, M.P., Achim, C.L., Wiley, C.A., Major, E.O., Saito, K., & Markey, S.P. (1996). Human microglia convert l-tryptophan into the neurotoxin quinolinic acid. *The Biochemical Journal,* Vol.320 (Pt 2), pp. 595-597

Hill, K.E., Pigmans, M., Fujinami, R.S., & Rose, J.W. (1998). Gender variations in early Theiler's virus induced demyelinating disease: differential susceptibility and effects of IL-4, IL-10 and combined IL-4 with IL-10. *Journal of Neuroimmunology,* Vol.85, No.1, pp. 44-51

Hirsch, J.D., Beyer, C.F., Malkowitz, L., Beer, B., & Blume, A.J. (1989). Mitochondrial benzodiazepine receptors mediate inhibition of mitochondrial respiratory control. *Molecular Pharmacology,* Vol.35, No.1, pp. 157-163

Hirtz, D., Thurman, D.J., Gwinn-Hardy, K., Mohamed, M., Chaudhuri, A.R., & Zalutsky, R. (2007). How common are the "common" neurologic disorders?. *Neurology,* Vol.68, No.5, pp. 326-337

Hung, Y.Y., & Huang, T.L. (2007). Lorazepam and diazepam for relieving catatonic features in multiple sclerosis. *Progress in Neuropsychopharmacology & Biological Psychiatry,* Vol.31, No.7, pp. 1537-1538

Inoue, A., Koh, C.S., Shimada, K., Yanagisawa, N., & Yoshimura, K. (1996). Suppression of cell-transferred experimental autoimmune encephalomyelitis in defibrinated Lewis rats. *Journal of Neuroimmunology,* Vol.71, No.1-2, pp. 131-137

Izikson, L., Klein, R.S., Charo, I.F., Weiner, H.L., & Luster, A.D. (2000). Resistance to experimental autoimmune encephalomyelitis in mice lacking the CC chemokine receptor (CCR)2. *The Journal of Experimental Medicine,* Vol.192, No.7, pp. 1075-80

Jayakumar, A.R., Panickar, K.S., & Norenberg, M.D. (2002). Effects on free radical generation by ligands of the peripheral benzodiazepine receptor in cultured neural cells. *Journal of Neurochemistry,* Vol.83, No.5, pp. 1226-1234

Johnson, R.R., Storts, R., Welsh, T.H. Jr., Welsh, C.J., & Meagher, M.W. (2004). Social stress alters the severity of acute Theiler's virus infection. *Journal of Neuroimmunology,* Vol.148, No.1-2, pp. 74-85

Johnson, R.R., Prentice, T.W., Bridegam, P., Young, C.R., Steelman, A.J., Welsh, T.H., Welsh, C.J., & Meagher, M.W. (2006). Social stress alters the severity and onset of the chronic phase of Theiler's virus infection. *Journal of Neuroimmunology,* Vol.175, No.1-2, pp. 39-51

Johnston, C., Jiang, W., Chu, T., & Levine, B. (2001). Identification of genes involved in the host response to neurovirulent alphavirus infection. *Journal of Virology,* Vol.75, No.21, pp. 10431-10445

Juedes, A.E., Hjelmström, P., Bergman, C.M., Neild, A.L., & Ruddle, N.H. (2000). Kinetics and cellular origin of cytokines in the central nervous system: insight into mechanisms of myelin oligodendrocyte glycoprotein-induced experimental autoimmune encephalomyelitis. *Journal of Immunology,* Vol.164, No.1, pp. 419-426

Junck, L., Olson, J.M., Ciliax, B.J., Koeppe, R.A., Watkins, G.L., Jewett, D.M., McKeever, P.E., Wieland, D.M., Kilbourn, M.R., Starosta-Rubinstein, S., Mancini, W.R., Kuhl, D.E., Greenberg, H.S., & Young, A.B. (1989). PET imaging of human gliomas with ligands for the peripheral benzodiazepine binding site. *Annals of Neurology,* Vol.26, No.6, pp. 752-758

Kappel, C.A., Melvold, R.W., & Kim, B.S. (1990). Influence of sex on susceptibility in the Theiler's murine encephalomyelitis virus model for multiple sclerosis. *Journal of Neuroimmunology,* Vol.29, No.1-3, pp. 15-19

Karpus, W.J., & Kennedy, K.J. (1997). MIP-1alpha and MCP-1 differentially regulate acute and relapsing autoimmune encephalomyelitis as well as Th1/Th2 lymphocyte differentiation. *Journal of Leukocyte Biology,* Vol.62, No.5, pp. 681-687

Karpus, W.J., & Ransohoff, R.M. (1998). Chemokine regulation of experimental autoimmune encephalomyelitis: temporal and spatial expression patterns govern disease pathogenesis. *Journal of Neuroimmunology,* Vol.161, No.6, pp. 2667-2671

Kornecki, E., Ehrlich, Y.H., & Lenox, R.H. (1984). Platelet-activating factor-induced aggregation of human platelets specifically inhibited by triazolobenzodiazepines. *Science,* Vol.226, No.4681, pp. 1454-1456

Kornecki, E., Lenox, R.H., Hardwick, D.H., Bergdahl, J.A., & Ehrlich, Y.H. (1987). Interactions of the alkyl-ether-phospholipid, platelet activating factor (PAF) with platelets, neural cells, and the psychotropic drugs triazolobenzodiazepines. *Advances in Experimental Medicine and Biology,* Vol.221, pp. 477-488

Kreutzberg, G.W. (1996). Microglia: a sensor for pathological events in the CNS. *Trends in Neurosciences,* Vol.19, No.8, pp. 312-318

Kuhlmann, A.C., & Guilarte, T.R. (2000). Cellular and subcellular localization of peripheral benzodiazepine receptors after trimethyltin neurotoxicity. *Journal of Neurochemistry,* Vol.74, No.4, pp. 1694-1704

Kuroda, Y., Mori, T., & Hori, T. (1994). Restraint stress suppresses experimental allergic encephalomyelitis in Lewis rats. *Brain Research Bulletin,* Vol.34, No.1, pp. 15-17

Laban, O., Dimitrijević, M., von Hoersten, S., Marković, B.M., & Janković, B.D. (1995a). Experimental allergic encephalomyelitis in adult DA rats subjected to neonatal handling or gentling. *Brain Research,* Vol.676, No.1, pp. 133-140

Laban, O., Marković, B.M., Dimitrijević, M., & Janković, B.D. (1995b). Maternal deprivation and early weaning modulate experimental allergic encephalomyelitis in the rat. *Brain, Behavior, and Immunity,* Vol.9, No.1, pp. 9-19

Lacapère, J.J., & Papadopoulos, V. (2003). Peripheral-type benzodiazepine receptor: structure and function of a cholesterol-binding protein in steroid and bile acid biosynthesis. *Steroids,* Vol.68, No.7-8, pp. 569-585

Lacor, P., Gandolfo, P., Tonon, M.C., Brault, E., Dalibert, I., Schumacher, M., Benavides, J., & Ferzaz, B. (1999). Regulation of the expression of peripheral benzodiazepine receptors and their endogenous ligands during rat sciatic nerve degeneration and regeneration: a role for PBR in neurosteroidogenesis. *Brain Research,* Vol.815, No.1, pp. 70-80

Lavi, E., & Constantinescu, C.S. (2005). *Experimental Models of Multiple Sclerosis,* Springer Verlag, ISBN 0-387-25517-6, New York

Le Goascogne, C., Eychenne, B., Tonon, M.C., Lachapelle, F., Baumann, N., & Robel, P. (2000). Neurosteroid progesterone is up-regulated in the brain of jimpy and shiverer mice. *Glia,* Vol.29, No.1, pp. 14-24

Le Page, C., Ferry, A., & Rieu, M. (1994). Effect of muscular exercise on chronic relapsing experimental autoimmune encephalomyelitis. *Journal of Applied Physiology,* Vol.77, No.5, pp. 2341-2347

Le Page, C., Bourdoulous, S., Béraud, E., Couraud, P.O., Rieu, M., & Ferry, A. (1996). Effect of physical exercise on adoptive experimental auto-immune encephalomyelitis in rats. *European Journal of Applied Physiology and Occupational Physiology,* Vol.73, No.1-2, pp. 130-135

Lenfant, M., Haumont, J., & Zavala, F. (1986). In vivo immunomodulating activity of PK 1195, a structurally unrelated ligand for "peripheral" benzodiazepine binding sites--I. Potentiation in mice of the humoral response to sheep red blood cells. *International Journal of Immunopharmacology,* Vol.8, No.7, pp. 825-828

Levine, S., & Saltzman, A. (1987). Nonspecific stress prevents relapses of experimental allergic encephalomyelitis in rats. *Brain, Behavior, and Immunity,* Vol.1, No.4, pp. 336-341

Levine, S., Strebel, R., Wenk, E.J., & Harman, P.J. (1962). Suppression of experimental allergic encephalomyelitis by stress. *Proceedings of the Society for Experimental Biology and Medicine,* Vol.109, pp. 294-298

Li, J., Johansen, C., Brønnum-Hansen, H., Stenager, E., Koch-Henriksen, N., & Olsen, J. (2004). The risk of multiple sclerosis in bereaved parents: A nationwide cohort study in Denmark. *Neurology,* Vol.62, No.5, pp. 726-729

Libbey, J.E., & Fujinami, R.S. (2011). Experimental autoimmune encephalomyelitis as a testing paradigm for adjuvants and vaccines. *Vaccine,* Vol.29, No.17, pp. 3356-3362

Lipton, H.L. (1975). Theiler's virus infection in mice: an unusual biphasic disease process leading to demyelination. *Infection and Immunity,* Vol.11, No.5, pp. 1147-1155

Loers, G., Aboul-Enein, F., Bartsch, U., Lassmann, H., & Schachner, M. (2004). Comparison of myelin, axon, lipid, and immunopathology in the central nervous system of differentially myelin-compromised mutant mice: a morphological and biochemical study. *Molecular and Cellular Neurosciences,* Vol.27, No.2, pp. 175-189

MacPhee, I.A., Antoni, F.A., & Mason, D.W. (1989). Spontaneous recovery of rats from experimental allergic encephalomyelitis is dependent on regulation of the immune system by endogenous adrenal corticosteroids. *The Journal of Experimental Medicine,* Vol.169, No.2, pp. 431-445

Maeda, J., Suhara, T., Zhang, M.R., Okauchi, T., Yasuno, F., Ikoma, Y., Inaji, M., Nagai, Y., Takano, A., Obayashi, S., & Suzuki, K. (2004). Novel peripheral benzodiazepine receptor ligand [11C]DAA1106 for PET: an imaging tool for glial cells in the brain. *Synapse,* Vol.52, No.4, pp. 283-291

Marino, F., Cattaneo, S., Cosentino, M., Rasini, E., Di Grazia, L., Fietta, A.M., Lecchini, S., & Frigo, G. (2001). Diazepam stimulates migration and phagocytosis of human neutrophils: possible contribution of peripheral-type benzodiazepine receptors and intracellular calcium. *Pharmacology,* Vol.63, No.1, pp. 42-49

Martinelli, V. (2000). Trauma, stress and multiple sclerosis. *Neurological Sciences,* Vol.21, No.4, pp. S849-852

Masood, A., Banerjee, B., Vijayan, V.K., & Ray, A. (2003). Modulation of stress-induced neurobehavioral changes by nitric oxide in rats. *European Journal of Pharmacology,* Vol.458, No.1-2, pp. 135-139

Matsushima, G.K., & Morell, P. (2001). The neurotoxicant, cuprizone, as a model to study demyelination and remyelination in the central nervous system. *Brain Pathology,* Vol.11, No.1, pp. 107-116

Mattner, F., Katsifis, A., Staykova, M., Ballantyne, P., & Willenborg, D.O. (2005). Evaluation of a radiolabelled peripheral benzodiazepine receptor ligand in the central nervous system inflammation of experimental autoimmune encephalomyelitis: a possible probe for imaging multiple sclerosis. *European Journal of Nuclear Medicine and Molecular Imaging,* Vol.32, No.5, pp. 557-563

McEnery, M.W., Snowman, A.M., Trifiletti, R.R., & Snyder, S.H. (1992). Isolation of the mitochondrial benzodiazepine receptor: association with the voltage-dependent anion channel and the adenine nucleotide carrier. *Proceedings of the National Academy of Sciences of the United States of America,* Vol.89, No.8, pp. 3170-3174

McManus, C.M., Brosnan, C.F., & Berman, J.W. (1998). Cytokine induction of MIP-1 alpha and MIP-1 beta in human fetal microglia. *Journal of Immunology,* Vol.160, No.3, pp. 1449-1455

McQualter, J.L., & Bernard, C.C. (2007). Multiple sclerosis: a battle between destruction and repair. *Journal of Neurochemistry,* Vol.100, No.2, pp. 295-306

Meagher, M.W., Johnson, R.R., Young, E.E., Vichaya, E.G., Lunt, S., Hardin, E.A., Connor, M.A., & Welsh, C.J. (2007). Interleukin-6 as a mechanism for the adverse effects of social stress on acute Theiler's virus infection. *Brain, Behavior, and Immunity,* Vol.21, No.8, pp. 1083-1095

Meythaler, J.M., Tuel, S.M., & Cross, L.L. (1991). Spinal cord seizures: a possible cause of isolated myoclonic activity in traumatic spinal cord injury: case report. *Paraplegia,* Vol.29, No.8, pp. 557-560

Mi, W., Belyavskyi, M., Johnson, R.R., Sieve, A.N., Storts, R., Meagher, M.W., & Welsh, C.J. (2004). Alterations in chemokine expression following Theiler's virus infection and restraint stress. *Journal of Neuroimmunology,* Vol.151, No.1-2, pp. 103-115

Mi, W., Prentice, T.W., Young, C.R., Johnson, R.R., Sieve, A.N., Meagher, M.W., & Welsh, C.J. (2006). Restraint stress decreases virus-induced pro-inflammatory cytokine mRNA expression during acute Theiler's virus infection. *Journal of Neuroimmunology,* Vol.178, No.1-2, pp. 49-61

Mohr, D.C., Goodkin, D.E., Bacchetti, P., Boudewyn, A.C., Huang, L., Marrietta, P., Cheuk, W., & Dee, B. (2000). Psychological stress and the subsequent appearance of new brain MRI lesions in MS. *Neurology,* Vol.55, No.1, pp. 55-61

Mohr, D.C., Hart, S.L., Julian, L., Cox, D., & Pelletier, D. (2004). Association between stressful life events and exacerbation in multiple sclerosis: a meta-analysis. *BMJ,* Vol.328, No.7442, pp. 731-735

Monje, M.L., Toda, H., & Palmer, T.D. (2003). Inflammatory blockade restores adult hippocampal neurogenesis. *Science,* Vol.302, No.5651, pp. 1760-1765

Monteyne, P., Bureau, J.F., & Brahic, M. (1997). The infection of mouse by Theiler's virus: from genetics to immunology. *Immunological Reviews,* Vol.159, No.1, pp. 163-176

Morgan, D., Gordon, M.N., Tan, J., Wilcock, D., & Rojiani, A.M. (2005). Dynamic complexity of the microglial activation response in transgenic models of amyloid deposition: implications for Alzheimer therapeutics. *Journal of Neuropathology and Experimental Neurology,* Vol.64, No.9, pp. 743-753

Murphy, G.M. Jr., Jia, X.C., Song, Y., Ong, E., Shrivastava, R., Bocchini, V., Lee, Y.L., & Eng, L.F. (1995). Macrophage inflammatory protein 1-alpha mRNA expression in an immortalized microglial cell line and cortical astrocyte cultures. *Journal of Neuroscience Research,* Vol.40, No.6, pp. 755-763

Ng, D.S., & Wong, K. (1988). Specific binding of platelet-activating factor (PAF) by human peripheral blood mononuclear leukocytes. *Biochemical and Biophysical Research Communications,* Vol.155, No.1, pp. 311-316

Nimmerjahn, A., Kirchhoff, F., & Helmchen, F. (2005). Resting microglial cells are highly dynamic surveillants of brain parenchyma in vivo. *Science,* Vol.308, No.5726, pp. 1314-1318

Nisipeanu, P., & Korczyn, A.D. (1993). Psychological stress as risk factor for exacerbations in multiple sclerosis. *Neurology,* Vol.43, No.7, pp. 1311-1312

Noseworthy, J.H., Lucchinetti, C., Rodriguez, M., & Weinshenker, B.G. (2000). Multiple sclerosis. *The New England Journal of Medicine,* Vol.343, No.13, pp. 938-952

Núñez, M.J., Riveiro, M.P., Varela, M., Liñares, D., Lopez, P., Maña, P., Martinez-Bahamonde, F., Núñez, L., Mayan, J.M., Rey-Mendez, M., & Freire-Garabal, M. (1998). Effects of nefazodone on delayed type hypersensitivity response in stressed mice. *Research Communications in Biological Psychology and Psychiatry,* Vol.23, No.1-2, pp. 19-27

Núñez, M.J., Balboa, J., Rodrigo, E., Brenlla, J., González-Peteiro, M., & Freire-Garabal, M. (2006). Effects of fluoxetine on cellular immune response in stressed mice. *Neuroscience Letters,* Vol.396, No.3, pp. 247-251

Núñez-Iglesias, M.J., Novío, S., Almeida-Dias, A., & Freire-Garabal, M. (2010). Inhibitory effects of alprazolam on the development of acute experimental autoimmune encephalomyelitis in stressed rats. *Pharmacology, Biochemistry, and Behavior,* Vol.97, No.2, pp. 350-356

Oda, T., Ueda, A., Shimizu, N., Handa, H., & Kasahara, T. (2002). Suppression of monocyte chemoattractant protein 1, but not IL-8, by alprazolam: effect of alprazolam on c-Rel/p65 and c-Rel/p50 binding to the monocyte chemoattractant protein 1 promoter region. *Journal of Immunology,* Vol.169, No.6, pp. 3329-3335

Oleszak, E.L., Chang, J.R., Friedman, H., Katsetos, C.D., & Platsoucas, C.D. (2004). Theiler's virus infection: a model for multiple sclerosis. *Clinical Microbiology Reviews,* Vol.17, No.1, pp. 174-207

Olson, J.M., Ciliax, B.J., Mancini, W.R., & Young, A.B. (1988). Presence of peripheral-type benzodiazepine binding sites on human erythrocyte membranes. *European Journal of Pharmacology,* Vol.152, No.1-2, pp. 47-53

Owens, M.J., Bissette, G., & Nemeroff, C.B. (1989). Acute effects of alprazolam and adinazolam on the concentrations of corticotropin-releasing factor in the rat brain. *Synapse,* Vol.4, No.3, pp. 196-202

Owens T. (2006). Animal models for multiple sclerosis. *Advances in Neurology,* Vol.98, No.1, pp. 77-89

Owhashi, M., Shouzui, Y., & Arita, H. (1997). Stress down-regulates experimental allergic encephalomyelitis (EAE) but permits activation and localization of autoreactive V beta 8.2+ T cells. *The International Journal of Neuroscience,* Vol.89, No.3-4, pp. 177-188

Pahan, K. (2010). Neuroimmune pharmacological control of EAE. *Journal of Neuroimmune Pharmacology,* Vol.5, No.2, pp. 165-168

Papadopoulos, V., Guarneri, P., Kreuger, K.E., Guidotti, A., & Costa, E. (1992). Pregnenolone biosynthesis in C6-2B glioma cell mitochondria: regulation by a mitochondrial diazepam binding inhibitor receptor. *Proceedings of the National Academy of Sciences of the United States of America,* Vol.89, No.11, pp. 5113-5117

Papadopoulos, V., Amri, H., Li, H., Boujrad, N., Vidic, B., & Garnier, M. (1997). Targeted disruption of the peripheral-type benzodiazepine receptor gene inhibits steroidogenesis in the R2C Leydig tumor cell line. *The Journal of Biological Chemistry,* Vol.272, No.51, pp. 32129-32135

Papadopoulos, V., Baraldi, M., Guilarte, T.R., Knudsen, T.B., Lacapère, J.J., Lindemann, P., Norenberg, M.D., Nutt, D., Weizman, A., Zhang, M.R., & Gavish, M. (2006a). Translocator protein (18kDa): new nomenclature for the peripheral-type benzodiazepine receptor based on its structure and molecular function. *Trends in Pharmacological Sciences,* Vol.27, No.8, pp. 402-409

Papadopoulos, V., Lecanu, L., Brown, R.C., Han, Z., & Yao, Z.X. (2006b). Peripheral-type benzodiazepine receptor in neurosteroid biosynthesis, neuropathology and neurological disorders. *Neuroscience,* Vol.138, No.3, pp. 749-756

Patte, C., Gandolfo, P., Leprince, J., Thoumas, J.L., Fontaine, M., Vaudry, H., & Tonon, M.C. (1999). GABA inhibits endozepine release from cultured rat astrocytes. *Glia,* Vol.25, No.4, pp. 404-411

Pawlikowski, M. (1993). Immunomodulating effects of peripherally acting benzodiazepines, In: *Peripheral benzodiazepine receptors,* E. Giesen-Crouse, (Ed), 125–135, Academic Press, London

Pérez-Nievas, B.G., García-Bueno, B., Madrigal, J.L., & Leza, J.C. (2010). Chronic immobilisation stress ameliorates clinical score and neuroinflammation in a MOG-induced EAE in Dark Agouti rats: mechanisms implicated. *Journal of Neuroinflammation,* Vol.7, pp. 60-72

Pratt, R.T. (1951). An investigation of the psychiatric aspects of disseminated sclerosis. *Journal of Neurology, Neurosurgery, and Psychiatry,* Vol.14, No.4, pp. 326-335

Ransohoff, R.M. (1999). Mechanisms of inflammation in MS tissue: adhesion molecules and chemokines. *Journal of Neuroimmunology,* Vol.98, No.1, pp. 57-68

Rey, C., Mauduit, C., Naureils, O., Benahmed, M., Louisot, P., & Gasnier, F. (2000). Up-regulation of mitochondrial peripheral benzodiazepine receptor expression by tumor necrosis factor alpha in testicular leydig cells. Possible involvement in cell survival. *Biochemical Pharmacology,* Vol.60, No.11, pp. 1639-1646

Righi, M., Mori, L., De Libero, G., Sironi, M., Biondi, A., Mantovani, A., Donini, S.D., & Ricciardi-Castagnoli, P. (1989). Monokine production by microglial cell clones. *European Journal of Immunology,* Vol.19, No.8, pp. 1443-1448

Rock, R.B., & Peterson, P.K. (2006). Microglia as a pharmacological target in infectious and inflammatory diseases of the brain. *Journal of Neuroimmune Pharmacology,* Vol.1, No.2, pp. 117-126

Rode, G., Maupas, E., Luaute, J., Courtois-Jacquin, S., & Boisson, D. (2003). [Medical treatment of spasticity]. *Neurochirurgie,* Vol.49, No.2-3, pp. 247–255

Rodriguez, M., Kenny, J.J., Thiemann, R.L., & Woloschak, G.E. (1990). Theiler's virus-induced demyelination in mice immunosuppressed with anti-IgM and in mice expressing the xid gene. *Microbial Pathogenesis,* Vol.8, No.1, pp. 23-35

Rothwell, N.J., & Hopkins, S.J. (1995). Cytokines and the nervous system II: Actions and mechanisms of action. *Trends in Neurosciences,* Vol.18, No.3, pp. 130-136

Ruff, M.R., Pert, C.B., Weber, R.J., Wahl, L.M., Wahl, S.M., & Paul, S.M. (1985). Benzodiazepine receptor-mediated chemotaxis of human monocytes. *Science,* Vol.229, No.4719, pp. 1281-1283

Ryu, J.K., Choi, H.B., & McLarnon, J.G. (2005). Peripheral benzodiazepine receptor ligand PK11195 reduces microglial activation and neuronal death in quinolinic acid-injected rat striatum. *Neurobiology of Disease,* Vol.20, No.2, pp. 550-561

Schwartz-Bloom, R.D., Miller, K.A., Evenson, D.A., Crain, B.J., & Nadler, J.V. (2000). Benzodiazepines protect hippocampal neurons from degeneration after transient cerebral ischemia: an ultrastructural study. *Neuroscience,* Vol.98, No.3, pp. 471-484

Sela, M. (2006). Immunomodulatory vaccines against autoimmune diseases. *Rejuvenation Research,* Vol.9, No.1, pp. 126-133

Selgas, L., Arce, A., Esquifino, A.I., & Cardinali, D.P. (1997). Twenty-four-hour rhythms of serum ACTH, prolactin, growth hormone, and thyroid-stimulating hormone, and of median-eminence norepinephrine, dopamine, and serotonin, in rats injected with Freund's adjuvant. *Chronobiology International,* Vol.14, No.3, pp. 253-265

Sibley, W.A. (1997). Risks factors in multiple sclerosis, In: *Multiple Sclerosis: Clinical and Pathogenic Basis,* C.S. Raine, H.F. McFarland, W.W. Tourtellotte, (Eds), 141-148, Chapman and Hall, London

Sieve, A.N., Steelman, A.J., Young, C.R., Storts, R., Welsh, T.H., Welsh, C.J., & Meagher, M.W. (2004). Chronic restraint stress during early Theiler's virus infection exacerbates the subsequent demyelinating disease in SJL mice. *Journal of Neuroimmunology,* Vol.155, No.1-2, pp. 103-118

Sieve, A.N., Steelman, A.J., Young, C.R., Storts, R., Welsh, T.H., Welsh, C.J., & Meagher, M.W. (2006). Sex-dependent effects of chronic restraint stress during early Theiler's virus infection on the subsequent demyelinating disease in CBA mice. *Journal of Neuroimmunology,* Vol.177, No.1-2, pp. 46-62

Simas, J.P., & Fazakerley, J.K. (1996). The course of disease and persistence of virus in the central nervous system varies between individual CBA mice infected with the BeAn strain of Theiler's murine encephalomyelitis virus. *The Journal of General Virology,* Vol.77, No.11, pp. 2701-2711

Solaro, C., & Messmer Uccelli, M. (2010). Pharmacological management of pain in patients with multiple sclerosis. *Drugs,* Vol.70, No.10, pp. 1245-1254

Sommer, N., Löschmann, P.A., Northoff, G.H., Weller, M., Steinbrecher, A., Steinbach, J.P., Lichtenfels, R., Meyermann, R., Riethmüller, A., Fontana, A., Dichgans, J., & Martin, R. (1995). The antidepressant rolipram suppresses cytokine production and prevents autoimmune encephalomyelitis. *Nature Medicine,* Vol.1, No.3, pp. 244-248

Sospedra, M., & Martin, R. (2005). Immunology of multiple sclerosis. *Annual Review of Immunology,* Vol.23, pp. 683-747

Steelman, A.J., Dean, D.D., Young, C.R., Smith, R. 3rd., Prentice, T.W., Meagher, M.W., & Welsh, C.J. (2009). Restraint stress modulates virus specific adaptive immunity during acute Theiler's virus infection. *Brain, Behavior, and Immunity,* Vol.23, No.6, pp. 830-843

Steelman, A.J., Alford, E., Young, C.R., Welsh, T.H., Meagher, M.W., & Welsh, C.J. (2010). Restraint stress fails to render C57BL/6 mice susceptible to Theiler's virus-induced demyelination. *Neuroimmunomodulation,* Vol.17, No.2, pp. 109-119

Steinman, L., & Zamvil, S.S. (2006). How to successfully apply animal studies in experimental allergic encephalomyelitis to research on multiple sclerosis. *Annals of Neurology,* Vol.60, No.1, pp. 12-21

Stoebner, P.E., Carayon, P., Casellas, P., Portier, M., Lavabre-Bertrand, T., Cuq, P., Cano, J.P., Meynadier, J., & Meunier, L. (2001). Transient protection by peripheral benzodiazepine receptors during the early events of ultraviolet light-induced apoptosis. *Cell Death and Differentiation,* Vol.8, No.7, pp. 747-753

Stork, C.M., & Hoffman, R.S. (1994). Characterization of 4-aminopyridine in overdose. *Journal of Toxicology. Clinical Toxicology,* Vol.32, No.5, pp. 583-587

Stüve, O., Youssef, S., Weber, M.S., Nessler, S., von Büdingen, H.C., Hemmer, B., Prod'homme, T., Sobel, R.A., Steinman, L., & Zamvil, S.S. (2006). Immunomodulatory synergy by combination of atorvastatin and glatiramer acetate in treatment of CNS autoimmunity. *The Journal of Clinical Investigation,* Vol.116,No.4, pp. 1037-1044

Teunis, M.A., Heijnen, C.J., Sluyter, F., Bakker, J.M., Van Dam, A.M., Hof, M., Cools, A.R., & Kavelaars, A. (2002). Maternal deprivation of rat pups increases clinical symptoms of experimental autoimmune encephalomyelitis at adult age. *Journal of Neuroimmunology,* Vol.133, No.1-2, pp. 30-38

Then Bergh, F., Kümpfel, T., Trenkwalder, C., Rupprecht, R., & Holsboer, F. (1999). Dysregulation of the hypothalamo-pituitary-adrenal axis is related to the clinical course of MS. *Neurology,* Vol.53, No.4, pp. 772-777

Tsunoda, I., & Fujinami, R.S. (1996). Two models for multiple sclerosis: experimental allergic encephalomyelitis and Theiler's murine encephalomyelitis virus. *Journal of Neuropathology and Experimental Neurology,* Vol.55, No.6, pp. 673-686

Varela-Patiño, M.P., Núñez, M.J., Losada, C., Pereiro, D., Castro-Bolaño, C., Saburido, X.L., Otero, J.M., Mayán, J.M., Blanco, A., Rey-Méndez, M., & Freire-Garabal, M. (1994). Effects of alprazolam on delayed type hypersensitivity (DTH) response in stressed mice. *Research Communications in Psychology, Psychiatry and Behavior,* Vol.19, No.1-2, pp. 69-77

Velez, L., Shirazi, F., Goto, C., Shepherd, G., & Roth, B.A. (2003). Opisthotonic posturing with neuromuscular irritability attributable to 4-aminopyridine ingestion by a healthy pediatric patient. *Pediatrics,* Vol.111, No.1, pp. e82-84

Venneti, S., Lopresti, B.J., & Wiley, C.A. (2006). The peripheral benzodiazepine receptor (Translocator protein 18kDa) in microglia: from pathology to imaging. *Progress in Neurobiology,* Vol.80, No.6, pp. 308-322

Versijpt, J., Debruyne, J.C., Van Laere, K.J., De Vos, F., Keppens, J., Strijckmans, K., Achten, E., Slegers, G., Dierckx, R.A., Korf, J., & De Reuck, J.L. (2005). Microglial imaging with positron emission tomography and atrophy measurements with magnetic resonance imaging in multiple sclerosis: a correlative study. *Multiple Sclerosis,* Vol.11, No.2, pp. 127-134

Vowinckel, E., Reutens, D., Becher, B., Verge, G., Evans, A., Owens, T., & Antel, J.P. (1997). PK11195 binding to the peripheral benzodiazepine receptor as a marker of microglia activation in multiple sclerosis and experimental autoimmune encephalomyelitis. *Journal of Neuroscience Research,* Vol.50, No.2, pp. 345-353

Warren, S., Greenhill, S., & Warren, K.G. (1982). Emotional stress and the development of multiple sclerosis: case-control evidence of a relationship. *Journal of Chronic Diseases,* Vol.35, No.11, pp. 821-831

Welsh, C.J., Tonks, P., Borrow, P., & Nash, A.A. (1990). Theiler's virus: an experimental model of virus-induced demyelination. *Autoimmunity,* Vol.6, No.1-2, pp. 105-112

Welsh, C.J., Bustamante, L., Nayak, M., Welsh, T.H., Dean, D.D., & Meagher, M.W. (2004). The effects of restraint stress on the neuropathogenesis of Theiler's virus infection II: NK cell function and cytokine levels in acute disease. *Brain, Behavior, and Immunity,* Vol.18, No.2, pp. 166-174

Welsh, C.J., Steelman, A.J., Mi, W., Young, C.R., Storts, R., Welsh, T.H. Jr., & Meagher, M.W. (2009). Neuroimmune interactions in a model of multiple sclerosis. *Annals of the New York Academy of Sciences,* Vol.1153, No.1, pp. 209-219

Whitacre, C.C., Dowdell, K., & Griffin, A.C. (1998). Neuroendocrine influences on experimental autoimmune encephalomyelitis. *Annals of the New York Academy of Sciences,* Vol.840, No.1, pp. 705-716

Whitacre, C.C., Reingold, S.C., & O'Looney, P.A. (1999). A gender gap in autoimmunity. *Science,* Vol.283, No.5406, pp. 1277-1278

Wilms, H., Claasen, J., Röhl, C., Sievers, J., Deuschl, G., & Lucius, R. (2003). Involvement of benzodiazepine receptors in neuroinflammatory and neurodegenerative diseases: evidence from activated microglial cells in vitro. *Neurobiology of Disease,* Vol.14, No.3, pp. 417-424

Woodruff, R.H., & Franklin, R.J. (1999). The expression of myelin protein mRNAs during remyelination of lysolecithin-induced demyelination. *Neuropathology and Applied Neurobiology,* Vol.25, No.3, pp. 226-235

Woods, M.J., & Williams, D.C. (1996). Multiple forms and locations for the peripheral-type benzodiazepine receptor. *Biochemical Pharmacology,* Vol.52, No.12, pp. 1805-1814

Yerdelen, D., Karatas, M., Goksel, B., & Yildirim, T. (2008). A patient with multiple sclerosis presenting with Holmes' tremor. *European Journal of Neurology,* Vol.15, No.1, pp. e2-3

Young, E.E., Prentice, T.W., Satterlee, D., McCullough, H., Sieve, A.N., Johnson, R.R., Welsh, T.H., Welsh, C.J., & Meagher, M.W. (2008). Glucocorticoid exposure alters the pathogenesis of Theiler's murine encephalomyelitis virus during acute infection. *Physiology & Behavior,* Vol.95, No.1-2, pp. 63-71

Young, E.E., Sieve, A.N., Vichaya, E.G., Carcoba, L.M., Young, C.R., Ambrus, A., Storts, R., Welsh, C.J., & Meagher, M.W. (2010). Chronic restraint stress during early Theiler's virus infection exacerbates the subsequent demyelinating disease in SJL mice: II. CNS disease severity. *Journal of Neuroimmunology,* Vol.220, No.1-2, pp. 79-89

Zavala, F., & Lenfant, M. (1987). Benzodiazepines and PK 11195 exert immunomodulating activities by binding on a specific receptor on macrophages. *Annals of the New York Academy of Sciences,* Vol.496, No.1, pp. 240-249

Zavala, F., Taupin, V., & Descamps-Latscha, B. (1990). In vivo treatment with benzodiazepines inhibits murine phagocyte oxidative metabolism and production of interleukin 1, tumor necrosis factor and interleukin-6. *The Journal of Pharmacology and Experimental Therapeutics,* Vol.255, No.2, pp. 442-450

7

Immunomodulation of Potent Antioxidant Agents: Preclinical Study to Clinical Application in Multiple Sclerosis

Shyi-Jou Chen[1,2] Hueng-Chuen Fan[1] and Huey-Kang Sytwu[2,*]
[1]*Department of Pediatrics, Tri-Service General Hospital*
[2]*Department of Microbiology and Immunology*
National Defense Medical Center
Taiwan (ROC)

1. Introduction

Multiple sclerosis (MS) is a chronic disease of the central nerve system (CNS) primarily affecting youngsters. The CNS has a potent antioxidant defense mechanism to scavenge reactive oxygen species (ROS). MS can be effectively studied in animal models with experimental autoimmune encephalomyelitis (EAE). The immunopathogenic mechanisms involved in EAE are attributed to T-cell-mediated inflammatory disease of the CNS and activation of T cells, which recruit invading inflammatory cells, such as macrophages. The activated T cells also induce astrocytes and microglia in situ to secrete cytokines, chemokines, and toxic molecules, namely, glutamate, nitric oxide (NO), and/or ROS, which in turn contribute to axonal damage. These processes are followed by complement activation and antibody-mediated phagocytosis of axons, which eventually lead to demyelination and axonal injury.

Brain tissue of MS patients was found to express antioxidants, and MS lesions have been reported to express high levels of antioxidants, indicating oxidative stress (OS). Heme oxygenase-1 (HO-1) is a heat-shock protein induced by OS; this protein was found to be expressed in active MS lesions and in EAE. Therefore, it can be inferred that ROS may play a distinct role in the pathogenesis of MS and that antioxidants may inhibit the development and progression of MS lesions. In this chapter, we summarize immunomodulation and therapeutic strategies in MS and examine experimental and clinical evidence to assess the applicability of the novel strategy of using antioxidants, such as melatonin, statins, α-lipoic acid (ALA), natural compounds (flavonoids), and erythropoietin (EPO), as adjuvant treatment in MS.

2. Roles of OS in MS

The initial stages of the formation of MS lesions are primarily characterized by the activation of residential microglia by autoreactive Th cells, while the active phase of the MS lesion is

* Corresponding Author
Conflict of interest: The authors declare that no conflict of interest exist

characterized by the infiltration of the lesion area by monocyte-derived macrophages, which initiate demyelination [Schreiner et al., 2009; van der Valk and De Groot, 2000]. Both activated microglia and infiltrated macrophages can generate excessive amounts of proinflammatory mediators and oxidizing radicals, such as superoxide, hydroxyl radicals, hydrogen peroxide, and NO [Colton and Gilbert, 1993; Gilgun-Sherki et al., 2004]. MS is most commonly studied in animal models of EAE. The immunopathogenesis of EAE is widely believed to involve T-cell-mediated inflammatory disease of the CNS, wherein activated T cells recruit macrophages and resident astrocytes and microglia, leading to the release of inflammatory mediators and cytotoxic molecules, namely, glutamate, NO, and/or ROS, which contribute to axonal damage; this is then followed by complement activation or antibody-mediated phagocytosis of axons [Hisahara et al., 2003; Schreibelt et al., 2007]. Furthermore, axonal degeneration in MS lesions can be divided into 2 steps: an initial stage of acute axonal injury in the inflammatory MS lesion [Ferguson et al., 1997; Trapp et al., 1998] and subsequent "slow burning" or axonal degeneration in non-inflammatory chronic lesions [Kornek et al., 2000]. The pathological changes in MS are also characterized by the presence of demyelinating plaques within the gray matter [Frohman et al., 2006]. Nevertheless, mitochondrial dysregulation and mitochondria-derived ROS have been reported to contribute to axonal damage in MS [Kalman et al., 2007; Mahad et al., 2009; Su et al., 2009; van Horssen et al., 2010; Witte et al., 2009]. In addition, a subset of activated microglia was found in cortical lesions, suggesting that microglia-derived ROS might contribute to gray matter demyelination [Gray et al., 2008]. Taken together, these evidences indicate that ROS play a pivotal role in several processes underlying the formation and persistence of MS lesions.

Experimental animal studies have demonstrated that the dietary intake of exogenous antioxidants, including flavonoids and α-lipoic acid, reduces the progression and clinical signs of EAE [Chaudhary et al., 2006; Hendriks et al., 2004; Marracci et al., 2002; Theoharides, 2009; Verbeek et al., 2005]. Despite promising results observed in animal models of MS, data on successful antioxidant therapy in MS patients is still limited; this emphasizes the need for epidemiological and clinical studies on this treatment strategy in MS. Reports indicate that antioxidants need be administered in high quantities to exert their protective effects in animal models of MS [Gilgun-Sherki et al., 2004; Mirshafiey and Mohsenzadegan, 2009]. Since ROS play a pivotal role in the initial phase as well as the chronic stage of MS, antioxidant therapy may be suitable for limiting overall disease progression. Thus, further understanding of the immunomodulatory activities of potential protective antioxidants is vital for their application in MS [Schreibelt et al., 2007; van Horssen et al.].

2.1 HO-1 in EAE/MS

HO-1 is a heat-shock protein induced by OS. In HO-1-expressing cells, the association of reductase with HO-1 competitively limits the interaction of reductase with cytochrome P450 isozymes, and thereby, the resultant production of superoxide; this reduced interaction limits free radical production and prevents oxidative damage of DNA, thereby suppressing oxidative or proinflammatory tissue damage [Prawan et al., 2005]. HO-1 has both antioxidative and anti-inflammatory activities and is highly inducible by a variety of stimuli, including its substrates heme and OS [Schipper, 2004]. Lee and Chau demonstrated that the overexpression of HO-1 in macrophages can inhibit proinflammatory response via lipopolysaccharide (LPS) stimulation and that IL-10 and HO-1 activate a positive feedback circuit to enhance the anti-inflammatory response both in vitro or in vivo [Lee and Chau, 2002]. In addition, Ponomarev et al. found that increased expression of IL-4 in glial cells was associated with reduced severity of EAE and that IL-4 production in the CNS is crucial for controlling autoimmune inflammation by inducing an alternative pathway for the regulation of microglial cells, but IL-4 production in the peripheral circulation was not found to have this effect [Ponomarev et al., 2007]. In contrast, Lee and Suk showed that IL-10 and IL-4 levels did not have any effect on the overexpression of HO-1 in LPS-stimulated microglia in vitro [Lee and Suk, 2007]. However, Zenclussen et al. found that the upregulation of HO-1 by using an adenoviral vector system expressing the HO-1 gene (AdHO-1) has a protective effect against fetal rejection in the murine abortion model; they also found that compared to abortion-prone mice, AdHO-1-treated mice showed higher values of the systemic and local IL-4/IFN-γ ratios and the IL-10/TNF-α ratio in the spleen [Zenclussen et al., 2006].

HO-1 has been detected in EAE lesions, and significant amplification of HO-1 protein levels has been proved in animal models of EAE [Schluesener and Seid, 2000]. HO-1 expression has also been reported in active MS lesions [van Horssen et al., 2008]. This enhanced expression of endogenous HO-1 may be one of the mechanisms involved in minimizing tissue damage in EAE [Emerson and LeVine, 2000]. Evidence has also shown that Hmox1(-/-)C57BL/6 mice, i.e., HO-1 gene knockout mice, displayed greater severity of EAE as compared to Hmox1(+/+) mice. Further, induction of HO-1 by cobalt protoporphyrin IX (CoPPIX) administration has been shown to suppress EAE progression, but this protective effect of CoPPIX was abrogated in Hmox1(-/-) mice with EAE [Chora et al., 2007]. Thus, endogenous HO-1 expression may play an important protective role in EAE, and therefore, the induction of HO-1 overexpression may represent a novel therapeutic strategy for MS [Liu et al., 2001].

2.2 Inflammation and immunopathogenesis of ROS and NO in MS

Brain tissue is very vulnerable to free radical damage because of its high oxygen utilization (20% of the total oxygen inspired); high concentrations of polyunsaturated fatty acids [Floyd and Hensley, 2002] and transition metals, such as iron, which are involved in the generation of the hydroxyl radical [Hill and Switzer, 1984]; and low concentrations of cytosolic antioxidants [Floyd and Carney, 1993; Reiter, 1995b]. In the brain, NO plays crucial roles in neuromodulation, neurotransmission, maintenance of synaptic plasticity, etc., and it also mediates pathological processes such as neurodegeneration and neuroinflammation [Golde et al., 2002]. Thus, NO may inhibit neuronal respiration, and NO production by astrocytes is believed to contribute to the neurodegenerative process via the impairment of mitochondrial function.

In addition, neuronal injury has also been reported to be associated with NO released by glial cells [Bal-Price and Brown, 2001; Mander et al., 2005]. Although glial activation can be protective, excess activation can be detrimental [Murphy, 1999]. Glia are activated by inflammatory mediators and express new proteins such as inducible NO synthatase (iNOS) [Emerit et al., 2004]. NO produced by the action of iNOS appears to be a key mediator of glia-induced neuronal death. Astroglial cells are "activated" in a wide range of CNS disorders, leading to the induction of iNOS [Bolanos et al., 1997; Endoh et al., 1994]. In mouse models of EAE, immunohistochemistry revealed elevated activity levels of iNOS in inflammatory lesions of the CNS, suggesting that excessive NO production due to activity of iNOS may play an essential role in eradicating inflammatory cells in the CNS of mice with EAE [Okuda et al., 1997].

A chronic inflammatory state gives rise to an activated immune response, which involves an acute phase protein response and the release of proinflammatory cytokines, macrophages, lymphocytes, and other immune system cells [Murakami, 2009; Taupin, 2008]. This complicated process triggers the recruitment of innate immune cells, which in turn mediates demyelination and axonal damage, and the formation of lesions that generally consist of T cells, macrophages, and microglia. OS plays a critical role in the pathogenesis of MS. ROS, which are generated in excess amounts mainly by macrophages subjected to OS, have been implicated in the demyelination and axonal damage occurring in both MS and EAE [Gilgun-Sherki et al., 2004]. NO has also been identified in the spinal cords of mouse models of EAE, indicating the potential role of NO in the pathogenesis of EAE, and possibly MS [Lin et al., 1993].

Increased levels of the indicators of OS and/or decreased levels of antioxidant enzymes and antioxidants have been detected in the blood and cerebrospinal fluid of MS patients in the active phases of the disease; these findings indicate that increased levels of ROS may cause the depletion of cellular antioxidants [van der Goes et al., 1998; van Horssen et al., 2010; van Meeteren et al., 2005]. In recent decades, immunologic cascade and inflammation have been proposed as causative factors of neurological diseases and autoimmune disorders, such as MS [Taupin, 2008]. On the basis of currently available findings, we assessed the usefulness of statins, melatonin, ALA, natural antioxidant compounds, and EPO, as potential candidates for adjuvant therapy in MS patients, by virtue of for their antioxidant and immunomodulatory proprieties.

3. Statins in MS/EAE

The 3-hydroxy-3-methylglutaryl coenzyme A (HMG-CoA) reductase inhibitors (or statins) are powerful cholesterol-lowering drugs, which are beneficial to the primary and secondary prevention of coronary heart disease; these drugs can improve endothelial function, increase NO bioavailability, exhibit antioxidant activity, stabilize atherosclerotic plaques, regulate progenitor cells, inhibit inflammatory responses, and exert immunomodulatory effects [Endres, 2006]. Pahan et al. reported that lovastatin inhibits the induction of iNOS and expression of proinflammatory cytokines in rat primary glial cells (astroglia and microglia) and macrophages [Pahan et al., 1997]. Subsequently, Stanislaus et al. showed that proinflammatory cytokines and iNOS are involved in the pathogenesis of EAE [Stanislaus et al., 1999] and thus highlighted the therapeutic importance of lovastatin in inhibiting the neuroinflammatory processes in the CNS and the central expression of iNOS, TNF-α, and IFN-γ in EAE; this suggests that lovastatin may have therapeutic potential in the treatment of neuroinflammatory diseases, such as MS [Stanislaus et al., 1999].

Atorvastatin exhibits pleiotropic immunomodulatory activity against both antigen-presenting cells (APC) and T-cell compartments. For instance, Youssef et al. reported that atorvastatin treatment of microglia inhibits IFN-γ-inducible transcription of multiple major histocompatibility complex (MHC) class II transactivator (CIITA) promoters and suppresses the upregulation of class II MHC [Youssef et al., 2002]. In addition, they found that atorvastatin suppresses IFN-γ-inducible expression of CD40, CD80, and CD86 co-stimulatory molecules, as well as L-mevalonate, thereby indicating that statins may be beneficial in the treatment of MS and other Th1-mediated autoimmune diseases [Youssef et al., 2002].

Nath et al. demonstrated that lovastatin inhibits EAE by modulating T-cell as well as APC responses and that the effects of lovastatin on macrophage/microglia in an Ag-non-specific system suggest that lovastatin is not only effective in EAE. Their study also revealed that lovastatin inhibits the transcription factors T-bet, NF-κB, and STAT4, which are responsible for CNS inflammation, via the induction of Th1 cell differentiation and production of related cytokines, such as IFN-γ and TNF-α. In addition, lovastatin induces the expression of GATA3 and STAT6 in Th2 cells and contributes to the downregulation of IFN-γ/T-bet in Th1 cells [Nath et al., 2004]. Statins have been reported to have potential for use as novel therapeutic agents to reverse the established paralysis in MS and to exert beneficial effects in synergy with other agents already approved for MS therapy [Weber and Zamvil, 2008]. However, simvastatin administered along with interferon β-1a as add-on therapy for at least 1–3 years in patients with relapsing–remitting MS (SIMCOMBIN study) did not show any beneficial effects in a placebo-controlled randomized phase 4 clinical trial [Sorensen et al.]. Similarly, a cohort study revealed that the disability progression in MS did not differ significantly between MS patients receiving and those not receiving statin therapy, thereby suggesting that statins do not affect the long-term course of MS [Paz Soldan et al.]. However, the combination of other statins with other disease-modifying drugs may be beneficial in MS. More clinical data is required to clarify this issue.

4. Melatonin in MS/EAE

Melatonin (5-methoxy-*N*-acetylserotonin) is mainly produced in the pineal gland during the dark phase of the day-night cycle, and it displays multifunctional properties and characteristics of a potent antioxidant [Martin et al., 2000]. In addition, melatonin is involved in the regulation of aging [Pierpaoli and Regelson, 1994] and scavenging of free hydroxyl radicals [Reiter, 1995a]. Melatonin is now well-known as a powerful antioxidant; an increasing number of experimental evidences have shown its protective effects against OS-induced macromolecular damage and diseases, including those involving mitochondrial dysfunction [Acuna et al., 2002]. Furthermore, melatonin exhibits both direct and indirect antioxidant activity, scavenges free radicals, stimulates antioxidant enzymes, enhances the activities of other antioxidants, and protects other antioxidant enzymes from oxidative damage [Castroviejo et al., 2011; Esposito and Cuzzocrea, 2010].

The results of numerous clinical studies have indicated that melatonin is a neuroprotective molecule in neurodegenerative disorders, which are believed to involve widespread brain oxidative damage [Esposito and Cuzzocrea, 2010; Kaur and Ling, 2008]. For example, melatonin defeats neurally derived free radicals and reduces the associated neuromorphological and neurobehavioral damage [Reiter et al., 2007]. Melatonin inhibits the expression of the iNOS in murine macrophages by inhibiting NF-κB activation [Gilad et al., 1998] and suppressing the levels of intercellular adhesion molecule (ICAM)-1 in experimental spinal cord reperfusion injury [Cuzzocrea et al., 2000]. Constantinescu et al reported that melatonin is capable of immunomodulation via the activation of NK cells [Constantinescu et al., 1997], upregulation of Th2 response, and inhibition of NF-κB [Maestroni, 1995]. Thus, both experimental and clinical data show that melatonin reduces the expression of adhesion molecules and pro-inflammatory cytokines and modifies serum inflammatory parameters.

Since the abovementioned findings indicate that melatonin exhibits both anti-inflammatory and immunomodulatory activities, it can be considered effective in suppressing autoimmune diseases, including EAE [Constantinescu, 1995; Sandyk, 1997]. Moreover, Kang et al. demonstrated that melatonin ameliorates EAE through the suppression of ICAM-1 levels. They also reported that melatonin treatment increases T-cell proliferation in mice and enhances the production of NK cells and monocytes in the bone marrow of mice, suggesting that it plays an important role in the immune system; thus, exogenous melatonin has been shown to ameliorate EAE by reducing the expression of ICAM-1 and lymphocyte function-associated antigen-1a (LFA-1α) in autoimmune target organs [Kang et al., 2001]. On the other hand, Maestroni et al. showed that melatonin treatment of mice decreases the expression of IL-2 and IFN-γ and upregulates the expression of Th2 cell cytokines, such as IL-4 and IL-10 [Maestroni, 1995]. Lin et al. found that melatonin prolongs islet graft survival in nonobese diabetic (NOD) mice via the reduction of Th1 cell and T-cell proliferation and elevating IL-10 levels, thereby indicating that melatonin treatment suppresses autoimmune recurrence after graft implantation [Lin et al., 2009]. Thus, melatonin plays a role in immunomodulation by regulating cytokine production of immunocompetent cells. From these findings, melatonin may be considered effective in improving the clinical course of autoimmune inflammatory diseases.

A clinical study by Akpinar et al. revealed that nocturnal serum melatonin levels were associated with major depression in acute MS patients, suggesting that melatonin deficiency may contribute to the occurrence of depression in MS patients [Akpinar et al., 2008b]. Moreover, Anderson et al. demonstrated that the optimization of melatonin and vitamin D3 inhibits the effects of IL-18 on the symptoms and cell loss of MS, as well as microglia and T-cell activation [Anderson and Rodriguez, 2011]. In addition, they confirmed that valproate treatment may interact significantly with melatonin and vitamin D3 to inhibit seizures and other signs and symptoms of MS [Anderson and Rodriguez, 2011]. Thus, the understanding of the immunomodulatory and anti-oxidative activity of melatonin in EAE may enable the therapeutic application of this molecule in MS.

5. ALA in MS/EAE

ALA (or 1,2-dithiolane-3-pentanoic acid) is a naturally occurring dithiol compound synthesized enzymatically in the mitochondrion from octanoic acid (Curr. Med. Chem. 11 (2004) 1135-1146.); it is an essential cofactor of key mitochondrial enzymes that control glucose oxidation, such as pyruvate dehydrogenase and α-ketoglutarate dehydrogenase [Gohil et al., 1999]. ALA mainly acts as a natural antioxidant that scavenges ROS and regenerates or recycles endogenous antioxidants (Free Radic. Res. 20 (1994) 119-133.). ALA and its reduced dithiol form, dihydrolipoic acid (DHLA), are potent antioxidants in both fat- and water-soluble media. In addition, ALA reacts with superoxide and hydroxyl radicals, hypochlorous acid, peroxyl radicals, and singlet oxygen [Marangon et al., 1999]. ALA also recycles vitamins C and E and increases the intracellular glutathione concentration. The antioxidative effect of ALA may be partially mediated by the chelation of transition metals, modification of the redox status of thiol-containing proteins, and inhibition of redox-sensitive nuclear transcription factors [Abdul and Butterfield, 2007; Akpinar et al., 2008a; Moini et al., 2002]. Tirosh et al. have previously shown that LA-plus is a potent protector of neuronal cells against glutamate-induced cytotoxicity and associated oxidative damage [Tirosh et al., 1999]. Further, Cheng et al. reported that ALA protects the cardiovascular system against oxidative injury` [Cheng et al., 2006]. They also found that ALA significantly increased the expression levels of HO-1 and ROS production and increased HO activity in A10 cells and the resistance of A10 cells to hydrogen-peroxide-induced OS; this effect was blocked by *N*-acetyl-cysteine, which also inhibited ALA-induced activation of p44/42 mitogen-activated protein kinase (MAPK) and AP-1, HO-1 expression, and HO activity. These findings suggest that ALA induces HO-1 expression through the production of ROS and subsequent activation of the p44/42 MAPK pathway and AP-1 in vascular smooth muscle cells and that it increases the expression of HO-1, which is a critical cytoprotective molecule [Cheng et al., 2006].

van der Goes et al. reported that ROS appear to be involved in the regulation of the phagocytosis of myelin and that lipoic acid (LA), a non-specific scavenger of ROS, also decreased the phagocytosis of myelin by macrophages [van der Goes et al., 1998].

Marracci et al. have shown that ALA ameliorates EAE in SJL mice immunized with proteolipid protein (PLP) 139-151 peptide, resulting in minimal inflammation characterized by less demyelination and only mild axonal loss in the spinal cords; they also demonstrated a marked reduction in the expression of $CD3^+$ T cells and $CD11b^+$ monocyte/macrophage cells in the affected spinal cord. Further, ALA and its reduced form, DHLA, inhibited the activity of matrix metalloproteinase-9 (MMP-9) in a dose-dependent manner [Marracci et al., 2002]. This was associated with a reduction in the number of CNS-infiltrating T cells and macrophages as well as decreased demyelination. Morini et al. further tested ALA in a therapeutic protocol aimed at suppressing myelin oligodendrocyte glycoprotein (MOG)-EAE; they also found significant reduction of demyelination and inflammatory infiltration and in the number of MOG-specific T cells, resulting in decreased production of IFN-γ and IL-4; this suggests that ALA exerts immunosuppressive activity on both Th1 and Th2 cytokines [Morini et al., 2004].

Marracci et al. showed that both ALA and DHLA inhibited Jurkat cell migration and have different mechanisms for inhibiting MMP-9 activity. These data together with the finding that ALA can ameliorate relapsing EAE suggest that ALA merits investigation as a therapeutic agent for MS [Marracci et al., 2004]. Moreover, Schreibelt et al. reported that antioxidant LA dose-dependently prevented the clinical development of EAE in a rat model of MS, along with a decrease in the CNS infiltration of leukocytes, particularly monocytes, which may be reflected as reduced ability to cross the blood brain barrier (BBB) [Schreibelt et al., 2006].

Accordingly, oxidative injury is recognized as an important process in the pathogenesis of MS. The powerful antioxidant property of ALA may render it a good candidate for adjuvant treatment in EAE/MS. In addition, ALA is greatly effective in suppressing and treating EAE; this effect is mediated by the inhibition of T-cell trafficking into the spinal cord, possibly by inhibiting MMP. Thus, ALA impedes the development of EAE not only by serving as an antioxidant but also by affecting the migratory capacity of monocytes and by stabilizing the BBB; these features make ALA an attractive therapeutic agent for MS

6. Potential of natural food compounds in the treatment of EAE/MS

Aktas et al. reported that the green tea extract epigallocatechin-3-gallate (EGCG) reduced the clinical severity of EAE when administered at the initiation or after the onset of the condition; this effect of EGCG is believed to be mediated by the limiting of brain inflammation and reducing neuronal damage, which are mediated by NF-kappa B inhibition, abrogated proliferation and TNF-alpha production of encephalitogenic T cells, and direct blocking of the formation of neurotoxic ROS in neurons [Aktas et al., 2004]. Thus, natural green tea components that have antioxidant, anti-inflammatory, and neuroprotective activities, may be considered novel candidates for the treatment of MS.

6.1 Natural compounds repress MMP to protect EAE/MS

Hendriks et al. have shown that flavonoids, which are naturally occurring compounds of green tea, can influence myelin phagocytosis by macrophages in vitro. The flavonoids luteolin, quercetin (3,3'4',5,7-pentahydroxy flavone), and fisetin most significantly reduce the extent of invasion of myelin phagocytes, without affecting their viability; this implies that they may be capable of restricting the demyelination process involved in MS [Hendriks et al., 2003]. On similar lines, Theoharides found that luteolin and structurally similar flavonoids can inhibit EAE and suggested that an appropriate luteolin formulation that permits sufficient absorption and reduces its metabolism could be a useful adjuvant to IFN-β in the treatment of MS [Theoharides, 2009]. A study by Muthian et al. on SJL/J mouse models of EAE revealed that quercetin (QRC), a flavonoid phytoestrogen, ameliorates EAE in vivo by inhibiting IL-12 production and neural antigen-specific Th1 differentiation and by reducing the MMP-9/TIMP-1 ratio. They further reported that when QRC was used in combination with IFN-β, it has additive effects on the regulation of the levels of TNF-α and MMP-9 [Sternberg et al., 2008]. The enzymes gelatinases A (MMP-2) and B (MMP-9) are involved in the pathogenesis of MS [Bever and Rosenberg, 1999; Mandler et al., 2001]. Liuzzi et al. reported that the non-flavonoids resveratrol (RSV) and tyrosol/hydroxytyrosol (Oliplus), but not the flavonoids QRC and catechins [green tea extract (GTE)], dose-dependently inhibit the expression of MMP-2 and MMP-9 in LPS-activated primary rat astrocytes. In turn, the direct inhibition of MMP2 and MMP-9 was achieved completely by QRC and GTE, not by RSV, and only partially by oliplus in cell-free systems of MS sera. These results indicate that the flavonoids and non-flavonoids tested exert their inhibitory effect on MMPs, displaying different mechanisms of action, possibly related to their structure; this indicates that their combined use may represent a powerful tool for the downregulation of MMPs in the course of MS [Liuzzi et al.].

6.2 Fumarate linking to Nrf2 pathway to protect MS

The antioxidant-responsive element (ARE) is an enhancer element that triggers the transcription of a battery of genes encoding phase II detoxification enzymes [Rushmore et al., 1991; Rushmore and Pickett, 1990] and factors vital for neuronal survival [Lee et al.,

2005]. The ARE is activated through the binding of its transcription factor NF-E2-related factor 2 (Nrf2) [Moi et al., 1994; Venugopal and Jaiswal, 1996]. Nrf2-ARE activation is a critical neuroprotective pathway that confers resistance to a variety of oxidative, stress-related, neurodegenerative injuries [Johnson et al., 2008]. Evidentially, autopsy studies of MS-affected tissue have revealed that neuronal Nrf2 is also activated during the natural course of MS, which is similar to the observations made in untreated EAE [Linker et al.]. Furthermore, van Horssen et al. showed that invading leukocytes contribute to cell damage and demyelination by producing excessive amounts of cytotoxic mediators, including ROS, and that to neutralize the destructive effects of ROS, the CNS is endowed with a repertoire of endogenous antioxidant enzymes, which are regulated by the transcription factor Nrf2; on the basis of these findings, they suggested that persistent Nrf2-mediated transcription occurs in active MS lesions, but this endogenous response is insufficient to prevent ROS-induced cellular damage, which is abundant in inflammatory MS lesions [van Horssen et al.]. Accordingly, overexpression of Nrf2/ARE-regulated antioxidants in EAE and MS tissue is indicative of ongoing OS [Schreibelt et al., 2007]. Furthermore, Linker et al. reported the Nrf2 pathway was activated by dimethyl fumarate or monomethyl fumarate [Linker et al.]. They showed that dimethylfumarate exerts protective effects on oligodendrocytes, myelin, axons, and neurons in vivo and reduces the oxidative stress in MOG-EAE models; their study also provided evidence that the proper functioning of Nrf2 is required for the therapeutic effect of dimethylfumarate, suggesting that the CNS-protective effects of dimethylfumarate involve the activation of Nrf2-mediated OS response mechanisms, which is an important protective mechanism of the CNS in a variety of pathological conditions. A phase II trial is currently underway to examine the applicability of fumarate as a disease-modifying agent in MS patients [Barten et al., 2010].

6.3 Other extracts from plants to benefit MS

De Paula et al. found that genistein, which occurs abundantly in soy products, has apoptotic, antioxidant, and anti-inflammatory properties. They observed that genistein treatment significantly ameliorated the severity of EAE, modulating pro- and anti-inflammatory cytokines and decreasing the rolling and adhering of leukocytes, which imply that genistein may have potential as a therapeutic agent for MS [De Paula et al., 2008]. Another antioxidant compound, silymarin—a purified extract from milk thistle (*Silybum marianum*)—is composed of a mixture of 4 isomeric flavonolignans: silibinin (main active component), isosilibinin, silydianin, and silychristin. This extract has been empirically used as a remedy for almost 2000 years and continues to be used in the treatment of many types of acute and chronic liver diseases. Although it is routinely used in clinical practice as a hepatoprotectant, the mechanisms underlying its beneficial effects remain largely unknown [Crocenzi and Roma, 2006]. Min et al. reported that silibinin significantly reduced the histological signs of demyelination and inflammation in EAE and that silibinin downregulated the secretion of pro-inflammatory Th1 cytokines and upregulated the anti-inflammatory Th2 cytokines in vitro in an Ag-nonspecific manner. Further, silibinin dose-dependently inhibited the production of Th1 cytokines ex vivo, indicating that it is both immunosuppressive and immunomodulatory and may therefore be effective in the treatment of MS [Min et al., 2007].

Together, many antioxidant components extracted from natural foods or plants, such as flavonoids, luteolin, EGCG (extracted from green tea), genistein (occurring abundantly in soy), and silibinin (the major pharmacologically active compound of silymarin, a fruit extract of *S. marianum*) have been proved to have not only health benefits but also the potential for use in the treatment of MS [Hutter and Laing, 1996; Mori et al., 2004; Sueoka et al., 2001; Theoharides, 2009].

7. Antioxidant and immunomodulatory effects of EPO in MS/EAE

EPO exhibits both hematopoietic and tissue-protective effects via interaction with different receptors [Leist et al., 2004].

Several experimental studies have shown that both EPO and EPO receptor (EpoR) are functionally expressed in the nervous system and that this cytokine has remarkable neuroprotective activity both in vitro, against different neurotoxicants, and in vivo, in animal models of experimental nervous system disorders [Bartesaghi et al., 2005]. Moreover, EPO has been proved to have the ability to cross the BBB and modulate astrocytes to protect the brain from ischemic damage and the spinal cord from injury in animal models [Bernaudin et al., 1999; Brines et al., 2000; Diaz et al., 2005]. The neuroprotective effects of EPO against neuronal death induced by ischemia and hypoxia have also been extensively studied in both in vitro and in vivo studies [Bernaudin et al., 1999; Gunnarson et al., 2009].

7.1 Cytoprotective interaction of EPO and HO-1

A recent report indicated that the upregulation of HO-1 expression contributes to EPO-mediated cytoprotection against myocardial ischemia–reperfusion injury [Burger et al., 2009]. The neuroprotective action of EPO in ischemic and CNS degenerative models was mediated by Janus-tyrosine kinase 2 (Jak2) signaling and the subsequent activation of PI3-K/Akt phosphorylation and NFκB cascades, which lead to the suppression of the CNS damage due to excitotoxins and the consequent generation of free radicals, including NO [Digicaylioglu and Lipton, 2001; Maiese et al., 2004]. Moreover, Sättler et al. showed that in rats with MOG-induced optic neuritis, the systemic administration of EPO resulted in a significant increase in the survival and function of retinal ganglion cells (RGC), the neurons that form the axons of the optic nerve; they also found that the neuroprotective effects of EPO were mediated by 3 independent intracellular signaling pathways involving the proteins phospho-Akt, phospho-MAPK 1 and 2, and Bcl-2m, which showed increased levels in vivo after EPO treatment [Sattler et al., 2004]. They also found that EPO in combination with a selective inhibitor of phosphatidylinositol 3-kinase (PI3-K) prevented the upregulation of phospho-Akt and the consecutive RGC rescue, thereby indicating that the PI3-K/Akt pathway in MOG-EAE has an essential influence on RGC survival under systemic EPO treatment [Sattler et al., 2004]. Interestingly, PI3K/Akt-pathway-related responses to OS and apoptosis have also been shown to be mediated by the transcriptional regulation of HO-1 [Martin et al., 2004]. Previously, Lifshitz et al. demonstrated that the dendritic cells (DCs) are direct targets of EPO, to initiate the immune response through the overexpression of human EPO in transgenic mice in in vivo experiments and validate a higher expression of EPO-R mRNA from bone marrow-derived DCs (BM-DCs) [Lifshitz et al., 2009]. Recent reports indicate that EPO exerts its cytoprotective effects in cardiac ischemia–reperfusion injury by inducing HO-1 expression [Burger et al., 2009]

7.2 Interaction of EPO, HO-1, and NO in EAE/MS

Tzima et al. reported that myeloid HO-1 deficiency exacerbated EAE in mice and enhanced the infiltration of activated macrophages and Th17 cells (IL-17-producing CD4+ IL-17-producing CD4+ T cells) in the CNS, and thus, they established HO-1 as a critical early mediator of the innate immune response [Tzima et al., 2009; Zwerina et al., 2005]. Liu showed that the inhibition of HO-1 expression resulted in marked exacerbation of EAE, suggesting that endogenous HO-1 expression plays an important protective role in EAE and that the targeted induction of HO-1 overexpression may represent a novel therapeutic strategy for the treatment of MS [Liu et al., 2001]. A study be Wu et al. at a center associated with ours revealed that the therapeutic induction of HO-1 expression ameliorates experimental murine membranous nephropathy via anti-oxidative, anti-apoptotic, and immunomodulatory mechanisms [Wu et al., 2008]. HO-1 overexpression has been proved to protect cells and tissues in a transgenic model of EAE [Panahian et al., 1999]. Moreover, Kumral et al. demonstrated that EPO exerts neuroprotective activity through the selective inhibition of NO overproduction in neonatal hypoxic–ischemic brain injury [Kumral et al., 2004]. Furthermore, Yuan et delineated a novel potential of EPO on peripheral inflammatory modulation in a murine MOG-EAE model [Yuan et al., 2008].

7.3 Immunomodulatory effects of EPO in EAE

After EPO was found to have neuroprotective effects in murine models of EAE [Zhang et al., 2005], it was introduced in humans and found to be effective in chronic progressive MS [Ehrenreich et al., 2007]. In our previous study, we found that EPO can enhance the expression of endogenous HO-1 either peripherally or locally in EAE; similarly, we observed that EPO-treated MOG-EAE mice exhibited upregulation of the splenic regulatory CD4+ T cells (Treg) and Th2 cells and downregulation of central Th1 and Th17 cells. We also obtained molecular evidence proving that EPO enhances the expression of endogenous HO-1 and that it has potential immunomodulatory activity and causes the suppression of inflammatory response to EAE [Chen et al., 2010].

We thus demonstrated the protective effects of EPO on EAE (Fig. 1a, b), and we observed significantly higher expression levels of endogenous HO-1 mRNA in the brain and a tendency for higher expression levels of endogenous HO-1 mRNA in the spinal cord and brain of the EPO-treated MOG-EAE mice than in those of the controls (Fig. 1c). Similarly, the expression levels of HO-1 mRNA in lymphocytes isolated from the CNS of MOG-EAE mice and controls differed significantly (Fig. 1c). Correspondingly, the protein levels of HO-1 in the spinal cord of EPO-treated MOG-EAE mice were higher than those of the controls (Fig. 1d). We further confirmed the augmenting effects of EPO on HO-1 in situ. Encephalitogenic Th17 cells play an essential role in the pathogenesis of EAE [Bettelli et al., 2006]. Th1 cells facilitate the invasion of Th17 cells to the CNS during EAE [O'Connor et al., 2008]. Further, IL-4 produced by CNS-derived Th2 cells is crucial to regulate the inflammation in EAE [Ponomarev et al., 2007]. To study the Th lineages further, we isolated mononuclear cells from the CNS of MOG-EAE mice treated with EPO and controls. Interestingly, we observed significantly lower ratios of both Th1 and Th17 cells to CD4+ cells in the EPO-treated MOG-EAE mice than in the controls; only a mildly increasing trend of encephalitogenic Th2 cells/CD4+ cells was noted in the EPO-treated group (Fig. 2 a-b). These findings suggest that EPO has the ability to counteract encephalitogenic Th1 and Th17 cells in situ, at least in part, and protect neuronal cells in EAE.

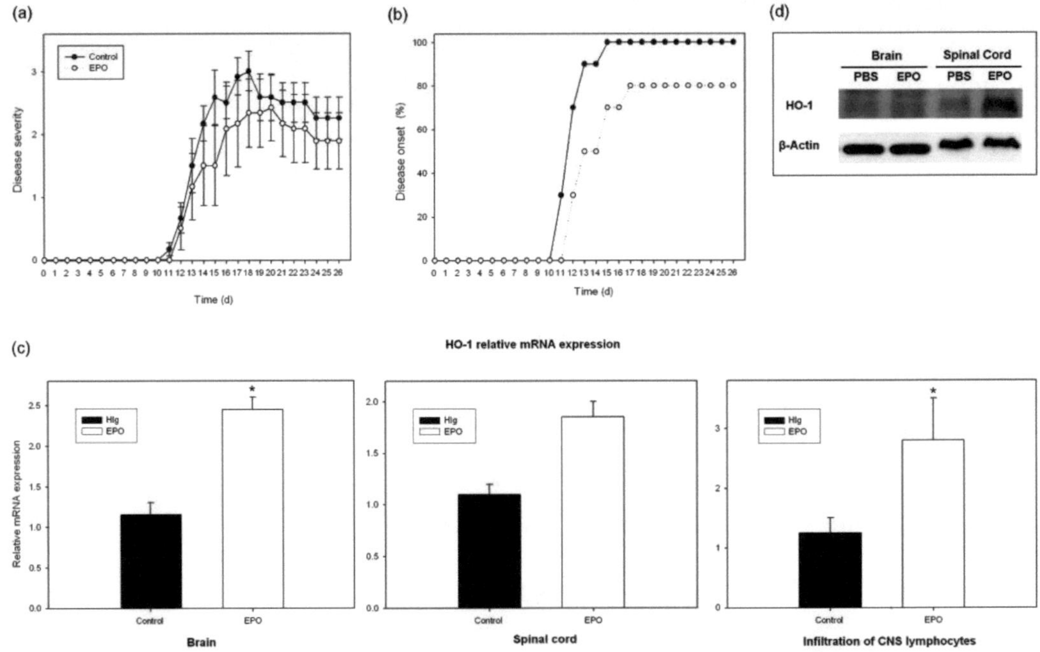

Fig. 1. EPO lessens EAE and enhances HO-1 *in situ*. (a) Clinical score and (b) time of onset of EAE in C57BL/6 mice treated i.p. with PBS or EPO (100 U/100 μl/mouse) on days 1, 3, 5, and 7 after s.c. immunization with MOG_{35-55}/CFA on day 0 and PTX i.p. on days 0 and 2. Each group contained 10 mice. Data represent means ± SEM. (c) Expression of HO-1 mRNA in the brain, spinal cord, or lymphocytes isolated from CNS of EPO-treated and PBS-treated (as control) MOG-EAE mice determined by real time PCR. Data in plots are expressed as mean ± SD from 3 independent experiments. Statistical significance was set at $p < 0.05$. (d) Western blot analysis of HO-1 expression in the brain and spinal cord of EPO-treated and control mice on day 14 after MOG injection. (partly adapted from Clin Exp Immunol. [Chen et al.])

Furthermore, we observed a greater extent of staining for HO-1-positive splenocytes in EPO-treated MOG-EAE mice than in the controls (Fig. 3a); this was also reflected in a significantly greater mean fluorescence intensity (MFI) in flow cytometry of the splenic lymphocytes of EPO-treated MOG-EAE mice than those of the controls (Fig. 3b). Similarly, we found an increased ratio of encephalitogenic Th1 and Th17 cells in EPO-treated MOG-EAE mice, however, only a mildly decreasing trend of splenic Th1 and Th17 cell subsets from EPO-treated MOG-EAE mice was noted. Instead, a significantly high ratio of splenic Th2 (Fig. 3a) was noted in the EPO-treated group than in the controls, and a notably significant elevation of splenic $CD25^{+}Foxp3^{+}$ $CD4^{+}$ cells (Tregs) was observed in the EPO-treated MOG-EAE mice (Fig. 3d).

We observed that the mRNA expression of HO-1 showed a tendency to increase, while the protein expression of HO-1 was notably high in the spinal cord of EPO-treated MOG-EAE; in contrast, a significant elevation of HO-1 mRNA, but no significant expression of HO-1 protein, was observed in the brain of EPO-treated MOG-EAE mice. Currently, Th17 cells are

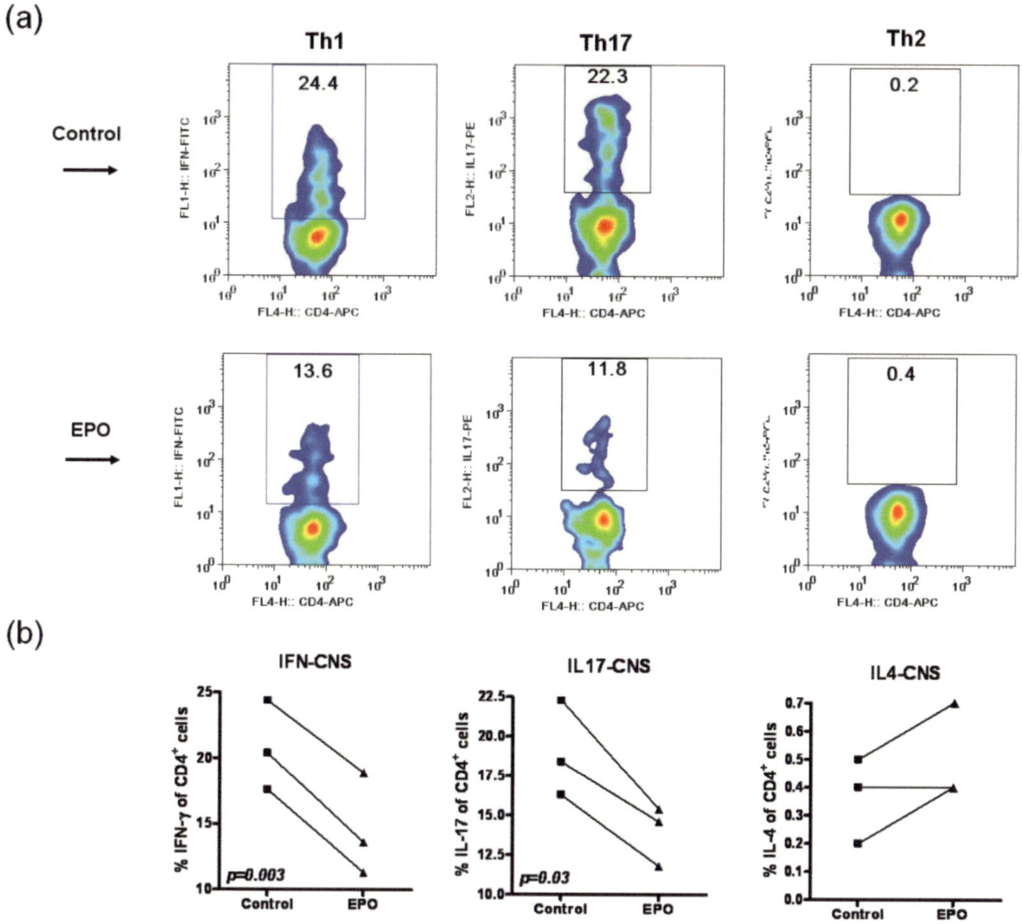

Fig. 2. Distribution of Th lymphocyte subsets in situ. (a) Flow cytometric analysis of intracellular cytokines in CD4+ T cells isolated from CNS of mice treated with either EPO or PBS. CD4 T lymphocytes isolated from the CNS of EPO-treated mice or controls on day 21 at peak disease stage were intracellularly stained with IL-4, IFN-γ, and IL-17 by flow cytometry. (b) Percentages of IFN-γ-producing CD4+ T cells (Th1), IL-17-producing CD4+ T cells (Th17), and IL-4-producing CD4+ T cells (Th2) are presented. Data are representative of 3 experiments. (partly adapted from Clin Exp Immunol. [Chen et al.])

believed to play a vital role in immunopathogenic mechanisms of EAE. However, Th1 cells are required to facilitate the CNS infiltration of Th17 cells in EAE [O'Connor et al., 2008]. Yuan et al. reported that short-tem EPO therapy for EAE can peripherally downregulate MHC class II of DCs and counteract Th17 cell responses [Yuan et al., 2008]. Our data confirmed further that EPO counteracts Th17 and Th1 cell inflammatory responses in EAE, both peripherally and centrally. EPO markedly reduced IL-6 levels in the spinal cord and decreased the inflammation and clinical score of EAE, thereby suggesting that the immunomodulatory activity of EPO may be partly mediated by the reduction of IL-6 levels

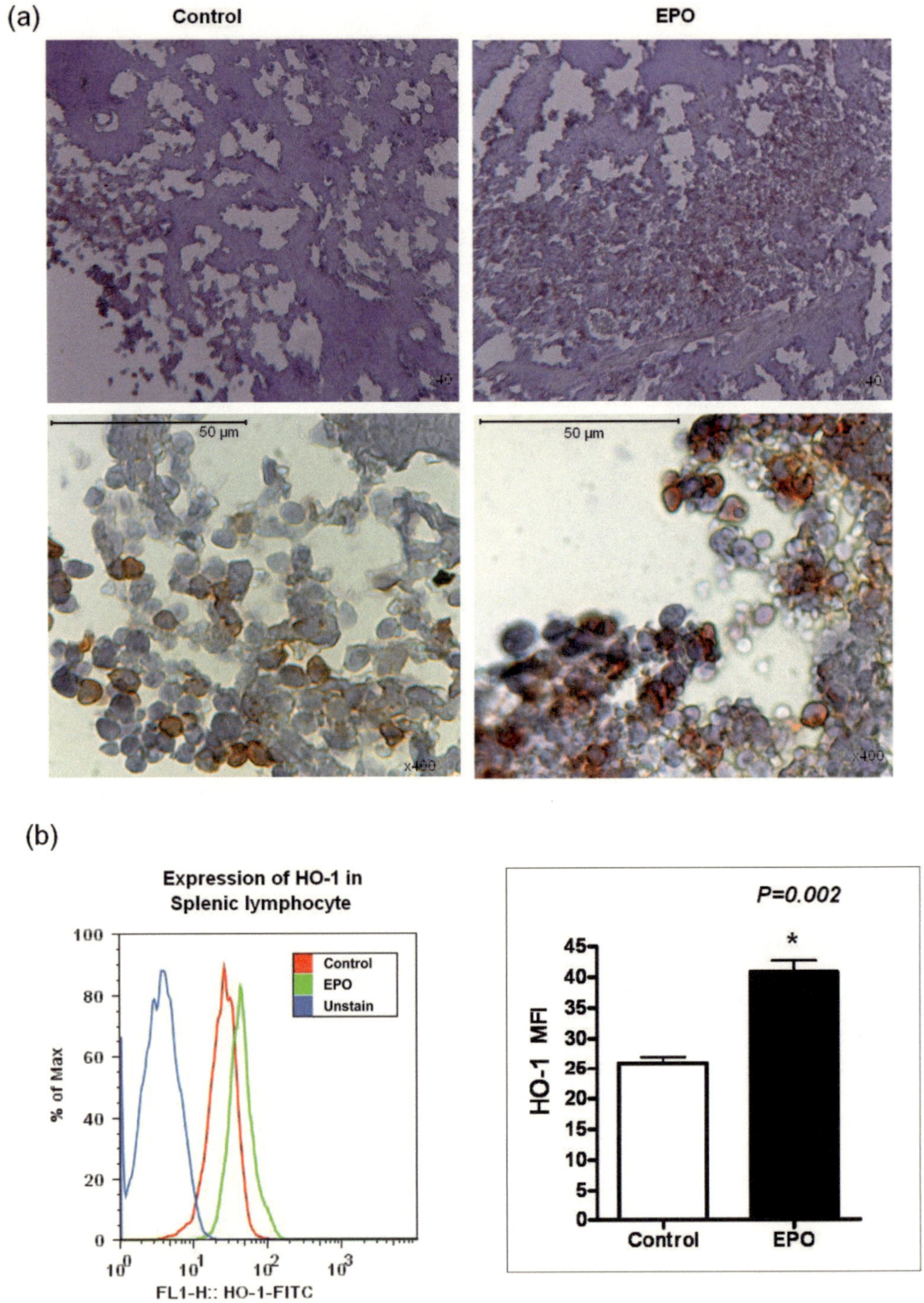
(a)
Control
EPO
x40
x40
50 μm
50 μm
x400
x400
(b)
Expression of HO-1 in Splenic lymphocyte
100
80
60
40
20
0
% of Max
Control
EPO
Unstain
10^0
10^1
10^2
10^3
FL1-H:: HO-1-FITC
P=0.002
*
45
40
35
30
25
20
15
10
5
0
HO-1 MFI
Control
EPO

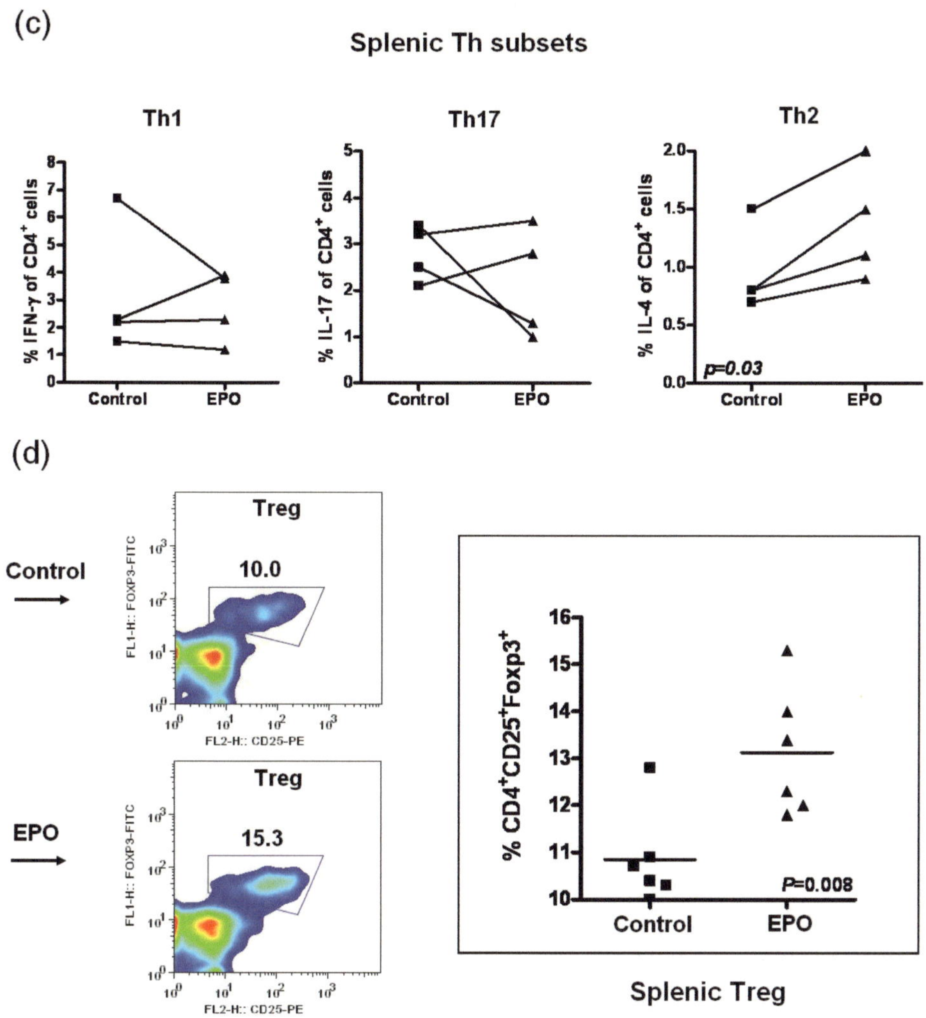

Fig. 3. EPO enhances splenic expression of HO-1, Th2 cells, and Treg.
(a) Immunohistochemical staining for HO-1 expression in the spleen of EPO-treated mice (right) and controls (left) on day 14 after MOG injection. Images are at either 40 × (top) or 400 × (bottom) magnification, and the length of the bar represents 50 μm. (b) Splenic lymphocytes from EPO-treated mice and controls on day 14 of EAE were stained with FITC-conjugated HO-1 for flow cytometry. Mean fluorescence intensity (MFI) of HO-1 staining was analyzed (b, left). Representative of 3 experiments. * indicates $p < 0.05$. (c) Splenic lymphocytes from EPO-treated mice and from controls were stained for Th1, Th2, and Th17 cells for flow cytometry analysis of the proportion of total CD4 T cells. Data represent 4 experiments. (d) These splenic lymphocytes were also stained for $CD25^+Foxp3^+$ $CD4^+$ cells (Treg) by flow cytometry. Compared to controls, EPO-treated MOG-EAE mice had higher ratio of splenic Treg on day 14. Data of Treg represent 6 experiments. (partly adapted from Clin Exp Immunol. [Chen et al.])

[Agnello et al., 2002]. Tron et al. noted that the expression of HO-1 were elevated in a localized inflammation after intramuscular injection of inflammatory material and IL-6-specific transcripts were introduced into the injured muscle and were in accordance with the serum levels of IL-6; these findings imply that the induction of HO-1 in local inflammation may affect other anti-inflammatory agents, such as local IL-6 [Tron et al., 2005]. Our data show that there is suppression of IL-6 mRNA in the CNS of EPO-treated MOG-EAE mice, which may be attributed to the overexpression of residential HO-1 to counteract IL-6 (Fig. 2). However, further investigation is required to clarify this.

We confirmed that exogenous EPO promotes the expression of endogenous HO-1, either in the CNS or the spleen, to repress Th1 and Th17 responses in situ and that it enhances the systemic invasion of Th2 and Tregs to reduce the severity of EAE. The potential role of EPO in upregulating the expression of endogenous HO-1 and, thereby, the anti-oxidative and anti-inflammatory activities of HO-1 indicate that EPO may have potential for clinical therapeutic application in autoimmune CNS disorders, such as MS. Taken together, these findings suggested that EPO not only causes the upregulation of HO-1 expression in the CNS but also acts as a potent inducer of HO-1 expression in the peripheral immunologic systems. Collectively, the neuroprotective action of EPO in EAE appears to involve different mechanisms of systemic and local inhibition of inflammation [Chen et al., 2010].

8. Conclusion

Patients with MS often experience difficulty in ambulation, spasticity, sensation, and cognition. The year 2010 marked the beginning of the era of oral medications for MS with the introduction of fingolimod [Brinkmann et al., 2010] as the first oral disease-modifying agent for MS. Subsequently other oral agents, including cladribine, teriflunomide, laquinimod, and dimethyl fumarate, as well as the monoclonal antibodies alemtuzumab, daclizumab, and rituximab have been used in MS [Gold, Krieger]. Currently, promising results have been obtained in Phase II trials of teriflunomide, daclizumab, laquinimod, and an antioxidant agent-fumarate [Barten et al., 2010].

However, to date, no specific drugs have been developed to completely cure MS.

Considering the data gathered from studies, such as those on animal models of EAE, it can be inferred that antioxidant molecules may be beneficial to some extent for adjuvant therapy in MS [Mirshafiey and Mohsenzadegan, 2009; van Horssen et al., 2008; van Horssen et al., 2010]. Although some antioxidants have shown some degree of efficacy in EAE, little information is available on the effect of their use in MS [Mirshafiey and Mohsenzadegan, 2009; Schreibelt et al., 2007]. Nevertheless, antioxidant therapy can be considered a candidate for use as adjuvant therapy in MS.

9. Acknowledgments

This work was supported by grants from the TSGH-C101-009-S03 and to S.J. Chen, National Science Council, Taiwan, Republic of China (NSC 99-2314-B-016 -002 -MY3 to S.J. Chen) and National Science Council, Taiwan, Republic of China (NSC100-3112-B-016-001 to H.-K. Sytwu).

10. References

Abdul HM, Butterfield DA (2007): Involvement of PI3K/PKG/ERK1/2 signaling pathways in cortical neurons to trigger protection by cotreatment of acetyl-L-carnitine and alpha-lipoic acid against HNE-mediated oxidative stress and neurotoxicity: implications for Alzheimer's disease. Free Radic Biol Med 42:371-84.

Acuna CD, Escames G, Carazo A, Leon J, Khaldy H, Reiter RJ (2002): Melatonin, mitochondrial homeostasis and mitochondrial-related diseases. Curr Top Med Chem 2:133-51.

Agnello D, Bigini P, Villa P, Mennini T, Cerami A, Brines ML, Ghezzi P (2002): Erythropoietin exerts an anti-inflammatory effect on the CNS in a model of experimental autoimmune encephalomyelitis. Brain Res 952:128-34.

Akpinar D, Yargicoglu P, Derin N, Aliciguzel Y, Agar A (2008a): The effect of lipoic acid on antioxidant status and lipid peroxidation in rats exposed to chronic restraint stress. Physiol Res 57:893-901.

Akpinar Z, Tokgoz S, Gokbel H, Okudan N, Uguz F, Yilmaz G (2008b): The association of nocturnal serum melatonin levels with major depression in patients with acute multiple sclerosis. Psychiatry Res 161:253-7.

Aktas O, Prozorovski T, Smorodchenko A, Savaskan NE, Lauster R, Kloetzel PM, Infante-Duarte C, Brocke S, Zipp F (2004): Green tea epigallocatechin-3-gallate mediates T cellular NF-kappa B inhibition and exerts neuroprotection in autoimmune encephalomyelitis. J Immunol 173:5794-800.

Anderson G, Rodriguez M (2011): Multiple sclerosis, seizures, and antiepileptics: role of IL-18, IDO, and melatonin. Eur J Neurol 18:680-5.

Bal-Price A, Brown GC (2001): Inflammatory neurodegeneration mediated by nitric oxide from activated glia-inhibiting neuronal respiration, causing glutamate release and excitotoxicity. J Neurosci 21:6480-91.

Barten LJ, Allington DR, Procacci KA, Rivey MP (2010): New approaches in the management of multiple sclerosis. Drug Des Devel Ther 4:343-66.

Bartesaghi S, Marinovich M, Corsini E, Galli CL, Viviani B (2005): Erythropoietin: a novel neuroprotective cytokine. Neurotoxicology 26:923-8.

Bernaudin M, Marti HH, Roussel S, Divoux D, Nouvelot A, MacKenzie ET, Petit E (1999): A potential role for erythropoietin in focal permanent cerebral ischemia in mice. J Cereb Blood Flow Metab 19:643-51.

Bettelli E, Carrier Y, Gao W, Korn T, Strom TB, Oukka M, Weiner HL, Kuchroo VK (2006): Reciprocal developmental pathways for the generation of pathogenic effector TH17 and regulatory T cells. Nature 441:235-8.

Bever CT, Jr., Rosenberg GA (1999): Matrix metalloproteinases in multiple sclerosis: targets of therapy or markers of injury? Neurology 53:1380-1.

Bolanos JP, Almeida A, Stewart V, Peuchen S, Land JM, Clark JB, Heales SJ (1997): Nitric oxide-mediated mitochondrial damage in the brain: mechanisms and implications for neurodegenerative diseases. J Neurochem 68:2227-40.

Brines ML, Ghezzi P, Keenan S, Agnello D, de Lanerolle NC, Cerami C, Itri LM, Cerami A (2000): Erythropoietin crosses the blood-brain barrier to protect against experimental brain injury. Proc Natl Acad Sci U S A 97:10526-31.

Brinkmann V, Billich A, Baumruker T, Heining P, Schmouder R, Francis G, Aradhye S, Burtin P (2010): Fingolimod (FTY720): discovery and development of an oral drug to treat multiple sclerosis. Nat Rev Drug Discov 9:883-97.

Burger D, Xiang F, Hammoud L, Lu X, Feng Q (2009): Role of heme oxygenase-1 in the cardioprotective effects of erythropoietin during myocardial ischemia and reperfusion. Am J Physiol Heart Circ Physiol 296:H84-93.

Castroviejo DA, Lopez LC, Escames G, Lopez A, Garcia JA, Reiter RJ (2011): Melatonin-mitochondria Interplay in Health and Disease. Curr Top Med Chem 11:221-40.

Chaudhary P, Marracci GH, Bourdette DN (2006): Lipoic acid inhibits expression of ICAM-1 and VCAM-1 by CNS endothelial cells and T cell migration into the spinal cord in experimental autoimmune encephalomyelitis. J Neuroimmunol 175:87-96.

Chen SJ, Wang YL, Lo WT, Wu CC, Hsieh CW, Huang CF, Lan YH, Wang CC, Chang DM, Sytwu HK (2010): Erythropoietin enhances endogenous haem oxygenase-1 and represses immune responses to ameliorate experimental autoimmune encephalomyelitis. Clin Exp Immunol 162:210-23.

Cheng PY, Lee YM, Shih NL, Chen YC, Yen MH (2006): Heme oxygenase-1 contributes to the cytoprotection of alpha-lipoic acid via activation of p44/42 mitogen-activated protein kinase in vascular smooth muscle cells. Free Radic Biol Med 40:1313-22.

Chora AA, Fontoura P, Cunha A, Pais TF, Cardoso S, Ho PP, Lee LY, Sobel RA, Steinman L, Soares MP (2007): Heme oxygenase-1 and carbon monoxide suppress autoimmune neuroinflammation. J Clin Invest 117:438-47.

Colton CA, Gilbert DL (1993): Microglia, an in vivo source of reactive oxygen species in the brain. Adv Neurol 59:321-6.

Constantinescu CS (1995): Melanin, melatonin, melanocyte-stimulating hormone, and the susceptibility to autoimmune demyelination: a rationale for light therapy in multiple sclerosis. Med Hypotheses 45:455-8.

Constantinescu CS, Hilliard B, Ventura E, Rostami A (1997): Luzindole, a melatonin receptor antagonist, suppresses experimental autoimmune encephalomyelitis. Pathobiology 65:190-4.

Crocenzi FA, Roma MG (2006): Silymarin as a new hepatoprotective agent in experimental cholestasis: new possibilities for an ancient medication. Curr Med Chem 13:1055-74.

Cuzzocrea S, Costantino G, Mazzon E, Micali A, De Sarro A, Caputi AP (2000): Beneficial effects of melatonin in a rat model of splanchnic artery occlusion and reperfusion. J Pineal Res 28:52-63.

De Paula ML, Rodrigues DH, Teixeira HC, Barsante MM, Souza MA, Ferreira AP (2008): Genistein down-modulates pro-inflammatory cytokines and reverses clinical signs of experimental autoimmune encephalomyelitis. Int Immunopharmacol 8:1291-7.

Diaz Z, Assaraf MI, Miller WH, Jr., Schipper HM (2005): Astroglial cytoprotection by erythropoietin pre-conditioning: implications for ischemic and degenerative CNS disorders. J Neurochem 93:392-402.

Digicaylioglu M, Lipton SA (2001): Erythropoietin-mediated neuroprotection involves cross-talk between Jak2 and NF-kappaB signalling cascades. Nature 412:641-7.

Ehrenreich H, Fischer B, Norra C, Schellenberger F, Stender N, Stiefel M, Siren AL, Paulus W, Nave KA, Gold R, Bartels C (2007): Exploring recombinant human erythropoietin in chronic progressive multiple sclerosis. Brain 130:2577-88.

Emerit J, Edeas M, Bricaire F (2004): Neurodegenerative diseases and oxidative stress. Biomed Pharmacother 58:39-46.

Emerson MR, LeVine SM (2000): Heme oxygenase-1 and NADPH cytochrome P450 reductase expression in experimental allergic encephalomyelitis: an expanded view of the stress response. J Neurochem 75:2555-62.

Endoh M, Maiese K, Wagner J (1994): Expression of the inducible form of nitric oxide synthase by reactive astrocytes after transient global ischemia. Brain Res 651:92-100.

Endres M (2006): Statins: potential new indications in inflammatory conditions. Atheroscler Suppl 7:31-5.

Esposito E, Cuzzocrea S (2010): Antiinflammatory activity of melatonin in central nervous system. Curr Neuropharmacol 8:228-42.

Ferguson B, Matyszak MK, Esiri MM, Perry VH (1997): Axonal damage in acute multiple sclerosis lesions. Brain 120 (Pt 3):393-9.

Floyd RA, Carney JM (1993): The role of metal ions in oxidative processes and aging. Toxicol Ind Health 9:197-214.

Floyd RA, Hensley K (2002): Oxidative stress in brain aging. Implications for therapeutics of neurodegenerative diseases. Neurobiol Aging 23:795-807.

Frohman EM, Racke MK, Raine CS (2006): Multiple sclerosis--the plaque and its pathogenesis. N Engl J Med 354:942-55.

Gilad E, Wong HR, Zingarelli B, Virag L, O'Connor M, Salzman AL, Szabo C (1998): Melatonin inhibits expression of the inducible isoform of nitric oxide synthase in murine macrophages: role of inhibition of NFkappaB activation. Faseb J 12:685-93.

Gilgun-Sherki Y, Melamed E, Offen D (2004): The role of oxidative stress in the pathogenesis of multiple sclerosis: the need for effective antioxidant therapy. J Neurol 251:261-8.

Gohil K, Roy S, Packer L, Sen CK (1999): Antioxidant regulation of gene expression: analysis of differentially expressed mRNAs. Methods Enzymol 300:402-10.

Gold R (2011): Oral therapies for multiple sclerosis: a review of agents in phase III development or recently approved. CNS Drugs 25:37-52.

Golde S, Chandran S, Brown GC, Compston A (2002): Different pathways for iNOS-mediated toxicity in vitro dependent on neuronal maturation and NMDA receptor expression. J Neurochem 82:269-82.

Gray E, Thomas TL, Betmouni S, Scolding N, Love S (2008): Elevated activity and microglial expression of myeloperoxidase in demyelinated cerebral cortex in multiple sclerosis. Brain Pathol 18:86-95.

Gunnarson E, Song Y, Kowalewski JM, Brismar H, Brines M, Cerami A, Andersson U, Zelenina M, Aperia A (2009): Erythropoietin modulation of astrocyte water permeability as a component of neuroprotection. Proc Natl Acad Sci U S A 106:1602-7.

Hendriks JJ, Alblas J, van der Pol SM, van Tol EA, Dijkstra CD, de Vries HE (2004): Flavonoids influence monocytic GTPase activity and are protective in experimental allergic encephalitis. J Exp Med 200:1667-72.

Hendriks JJ, de Vries HE, van der Pol SM, van den Berg TK, van Tol EA, Dijkstra CD (2003): Flavonoids inhibit myelin phagocytosis by macrophages; a structure-activity relationship study. Biochem Pharmacol 65:877-85.

Hill JM, Switzer RC, 3rd (1984): The regional distribution and cellular localization of iron in the rat brain. Neuroscience 11:595-603.

Hisahara S, Okano H, Miura M (2003): Caspase-mediated oligodendrocyte cell death in the pathogenesis of autoimmune demyelination. Neurosci Res 46:387-97.

Hutter CD, Laing P (1996): Multiple sclerosis: sunlight, diet, immunology and aetiology. Med Hypotheses 46:67-74.

Johnson JA, Johnson DA, Kraft AD, Calkins MJ, Jakel RJ, Vargas MR, Chen PC (2008): The Nrf2-ARE pathway: an indicator and modulator of oxidative stress in neurodegeneration. Ann N Y Acad Sci 1147:61-9.

Kalman B, Laitinen K, Komoly S (2007): The involvement of mitochondria in the pathogenesis of multiple sclerosis. J Neuroimmunol 188:1-12.
Kang JC, Ahn M, Kim YS, Moon C, Lee Y, Wie MB, Lee YJ, Shin T (2001): Melatonin ameliorates autoimmune encephalomyelitis through suppression of intercellular adhesion molecule-1. J Vet Sci 2:85-9.
Kaur C, Ling EA (2008): Antioxidants and neuroprotection in the adult and developing central nervous system. Curr Med Chem 15:3068-80.
Kornek B, Storch MK, Weissert R, Wallstroem E, Stefferl A, Olsson T, Linington C, Schmidbauer M, Lassmann H (2000): Multiple sclerosis and chronic autoimmune encephalomyelitis: a comparative quantitative study of axonal injury in active, inactive, and remyelinated lesions. Am J Pathol 157:267-76.
Krieger S (2011): Multiple sclerosis therapeutic pipeline: opportunities and challenges. Mt Sinai J Med 78:192-206.
Kumral A, Baskin H, Gokmen N, Yilmaz O, Genc K, Genc S, Tatli MM, Duman N, Ozer E, Ozkan H (2004): Selective inhibition of nitric oxide in hypoxic-ischemic brain model in newborn rats: is it an explanation for the protective role of erythropoietin? Biol Neonate 85:51-4.
Lee JM, Li J, Johnson DA, Stein TD, Kraft AD, Calkins MJ, Jakel RJ, Johnson JA (2005): Nrf2, a multi-organ protector? Faseb J 19:1061-6.
Lee S, Suk K (2007): Heme oxygenase-1 mediates cytoprotective effects of immunostimulation in microglia. Biochem Pharmacol 74:723-9.
Lee TS, Chau LY (2002): Heme oxygenase-1 mediates the anti-inflammatory effect of interleukin-10 in mice. Nat Med 8:240-6.
Leist M, Ghezzi P, Grasso G, Bianchi R, Villa P, Fratelli M, Savino C, Bianchi M, Nielsen J, Gerwien J, Kallunki P, Larsen AK, Helboe L, Christensen S, Pedersen LO, Nielsen M, Torup L, Sager T, Sfacteria A, Erbayraktar S, Erbayraktar Z, Gokmen N, Yilmaz O, Cerami-Hand C, Xie QW, Coleman T, Cerami A, Brines M (2004): Derivatives of erythropoietin that are tissue protective but not erythropoietic. Science 305:239-42.
Lifshitz L, Prutchi-Sagiv S, Avneon M, Gassmann M, Mittelman M, Neumann D (2009): Non-erythroid activities of erythropoietin: Functional effects on murine dendritic cells. Mol Immunol 46:713-21.
Lin GJ, Huang SH, Chen YW, Hueng DY, Chien MW, Chia WT, Chang DM, Sytwu HK (2009): Melatonin prolongs islet graft survival in diabetic NOD mice. J Pineal Res 47:284-92.
Lin RF, Lin TS, Tilton RG, Cross AH (1993): Nitric oxide localized to spinal cords of mice with experimental allergic encephalomyelitis: an electron paramagnetic resonance study. J Exp Med 178:643-8.
Linker RA, Lee DH, Ryan S, van Dam AM, Conrad R, Bista P, Zeng W, Hronowsky X, Buko A, Chollate S, Ellrichmann G, Bruck W, Dawson K, Goelz S, Wiese S, Scannevin RH, Lukashev M, Gold R (2011): Fumaric acid esters exert neuroprotective effects in neuroinflammation via activation of the Nrf2 antioxidant pathway. Brain 134:678-92.
Liu Y, Zhu B, Luo L, Li P, Paty DW, Cynader MS (2001): Heme oxygenase-1 plays an important protective role in experimental autoimmune encephalomyelitis. Neuroreport 12:1841-5.
Liuzzi GM, Latronico T, Brana MT, Gramegna P, Coniglio MG, Rossano R, Larocca M, Riccio P (2011): Structure-dependent inhibition of gelatinases by dietary

antioxidants in rat astrocytes and sera of multiple sclerosis patients. Neurochem Res 36:518-27.

Maestroni GJ (1995): T-helper-2 lymphocytes as a peripheral target of melatonin. J Pineal Res 18:84-9.

Mahad DJ, Ziabreva I, Campbell G, Lax N, White K, Hanson PS, Lassmann H, Turnbull DM (2009): Mitochondrial changes within axons in multiple sclerosis. Brain 132:1161-74.

Maiese K, Li F, Chong ZZ (2004): Erythropoietin in the brain: can the promise to protect be fulfilled? Trends Pharmacol Sci 25:577-83.

Mander P, Borutaite V, Moncada S, Brown GC (2005): Nitric oxide from inflammatory-activated glia synergizes with hypoxia to induce neuronal death. J Neurosci Res 79:208-15.

Mandler RN, Dencoff JD, Midani F, Ford CC, Ahmed W, Rosenberg GA (2001): Matrix metalloproteinases and tissue inhibitors of metalloproteinases in cerebrospinal fluid differ in multiple sclerosis and Devic's neuromyelitis optica. Brain 124:493-8.

Marangon K, Devaraj S, Tirosh O, Packer L, Jialal I (1999): Comparison of the effect of alpha-lipoic acid and alpha-tocopherol supplementation on measures of oxidative stress. Free Radic Biol Med 27:1114-21.

Marracci GH, Jones RE, McKeon GP, Bourdette DN (2002): Alpha lipoic acid inhibits T cell migration into the spinal cord and suppresses and treats experimental autoimmune encephalomyelitis. J Neuroimmunol 131:104-14.

Marracci GH, McKeon GP, Marquardt WE, Winter RW, Riscoe MK, Bourdette DN (2004): Alpha lipoic acid inhibits human T-cell migration: implications for multiple sclerosis. J Neurosci Res 78:362-70.

Martin D, Rojo AI, Salinas M, Diaz R, Gallardo G, Alam J, De Galarreta CM, Cuadrado A (2004): Regulation of heme oxygenase-1 expression through the phosphatidylinositol 3-kinase/Akt pathway and the Nrf2 transcription factor in response to the antioxidant phytochemical carnosol. J Biol Chem 279:8919-29.

Martin M, Macias M, Escames G, Leon J, Acuna-Castroviejo D (2000): Melatonin but not vitamins C and E maintains glutathione homeostasis in t-butyl hydroperoxide-induced mitochondrial oxidative stress. Faseb J 14:1677-9.

Min K, Yoon WK, Kim SK, Kim BH (2007): Immunosuppressive effect of silibinin in experimental autoimmune encephalomyelitis. Arch Pharm Res 30:1265-72.

Mirshafiey A, Mohsenzadegan M (2009): Antioxidant therapy in multiple sclerosis. Immunopharmacol Immunotoxicol 31:13-29.

Moi P, Chan K, Asunis I, Cao A, Kan YW (1994): Isolation of NF-E2-related factor 2 (Nrf2), a NF-E2-like basic leucine zipper transcriptional activator that binds to the tandem NF-E2/AP1 repeat of the beta-globin locus control region. Proc Natl Acad Sci U S A 91:9926-30.

Moini H, Packer L, Saris NE (2002): Antioxidant and prooxidant activities of alpha-lipoic acid and dihydrolipoic acid. Toxicol Appl Pharmacol 182:84-90.

Mori A, Yokoi I, Noda Y, Willmore LJ (2004): Natural antioxidants may prevent posttraumatic epilepsy: a proposal based on experimental animal studies. Acta Med Okayama 58:111-8.

Morini M, Roccatagliata L, Dell'Eva R, Pedemonte E, Furlan R, Minghelli S, Giunti D, Pfeffer U, Marchese M, Noonan D, Mancardi G, Albini A, Uccelli A (2004): Alpha-lipoic acid is effective in prevention and treatment of experimental autoimmune encephalomyelitis. J Neuroimmunol 148:146-53.

Murakami A (2009): Chemoprevention with phytochemicals targeting inducible nitric oxide synthase. Forum Nutr 61:193-203.

Murphy MP (1999): Nitric oxide and cell death. Biochim Biophys Acta 1411:401-14.

Nath N, Giri S, Prasad R, Singh AK, Singh I (2004): Potential targets of 3-hydroxy-3-methylglutaryl coenzyme A reductase inhibitor for multiple sclerosis therapy. J Immunol 172:1273-86.

O'Connor RA, Prendergast CT, Sabatos CA, Lau CW, Leech MD, Wraith DC, Anderton SM (2008): Cutting edge: Th1 cells facilitate the entry of Th17 cells to the central nervous system during experimental autoimmune encephalomyelitis. J Immunol 181:3750-4.

Okuda Y, Sakoda S, Fujimura H, Yanagihara T (1997): Nitric oxide via an inducible isoform of nitric oxide synthase is a possible factor to eliminate inflammatory cells from the central nervous system of mice with experimental allergic encephalomyelitis. J Neuroimmunol 73:107-16.

Pahan K, Sheikh FG, Namboodiri AM, Singh I (1997): Lovastatin and phenylacetate inhibit the induction of nitric oxide synthase and cytokines in rat primary astrocytes, microglia, and macrophages. J Clin Invest 100:2671-9.

Panahian N, Yoshiura M, Maines MD (1999): Overexpression of heme oxygenase-1 is neuroprotective in a model of permanent middle cerebral artery occlusion in transgenic mice. J Neurochem 72:1187-203.

Paz Soldan MM, Pittock SJ, Weigand SD, Yawn BP, Rodriguez M (2011): Statin therapy and multiple sclerosis disability in a population-based cohort. Mult Scler. Epub ahead of print.

Pierpaoli W, Regelson W (1994): Pineal control of aging: effect of melatonin and pineal grafting on aging mice. Proc Natl Acad Sci U S A 91:787-91.

Ponomarev ED, Maresz K, Tan Y, Dittel BN (2007): CNS-derived interleukin-4 is essential for the regulation of autoimmune inflammation and induces a state of alternative activation in microglial cells. J Neurosci 27:10714-21.

Prawan A, Kundu JK, Surh YJ (2005): Molecular basis of heme oxygenase-1 induction: implications for chemoprevention and chemoprotection. Antioxid Redox Signal 7:1688-703.

Reiter RJ (1995a): Functional pleiotropy of the neurohormone melatonin: antioxidant protection and neuroendocrine regulation. Front Neuroendocrinol 16:383-415.

Reiter RJ (1995b): Oxidative processes and antioxidative defense mechanisms in the aging brain. Faseb J 9:526-33.

Reiter RJ, Tan DX, Manchester LC, Tamura H (2007): Melatonin defeats neurally-derived free radicals and reduces the associated neuromorphological and neurobehavioral damage. J Physiol Pharmacol 58 Suppl 6:5-22.

Rushmore TH, Morton MR, Pickett CB (1991): The antioxidant responsive element. Activation by oxidative stress and identification of the DNA consensus sequence required for functional activity. J Biol Chem 266:11632-9.

Rushmore TH, Pickett CB (1990): Transcriptional regulation of the rat glutathione *S*-transferase Ya subunit gene. Characterization of a xenobiotic-responsive element controlling inducible expression by phenolic antioxidants. J Biol Chem 265:14648-53.

Sandyk R (1997): Influence of the pineal gland on the expression of experimental allergic encephalomyelitis: possible relationship to the acquisition of multiple sclerosis. Int J Neurosci 90:129-33.

Sattler MB, Merkler D, Maier K, Stadelmann C, Ehrenreich H, Bahr M, Diem R (2004): Neuroprotective effects and intracellular signaling pathways of erythropoietin in a rat model of multiple sclerosis. Cell Death Differ 11 Suppl 2:S181-92.

Schipper HM (2004): Heme oxygenase-1: transducer of pathological brain iron sequestration under oxidative stress. Ann N Y Acad Sci 1012:84-93.

Schluesener HJ, Seid K (2000): Heme oxygenase-1 in lesions of rat experimental autoimmune encephalomyelitis and neuritis. J Neuroimmunol 110:114-20.

Schreibelt G, Musters RJ, Reijerkerk A, de Groot LR, van der Pol SM, Hendrikx EM, Dopp ED, Dijkstra CD, Drukarch B, de Vries HE (2006): Lipoic acid affects cellular migration into the central nervous system and stabilizes blood-brain barrier integrity. J Immunol 177:2630-7.

Schreibelt G, van Horssen J, van Rossum S, Dijkstra CD, Drukarch B, de Vries HE (2007): Therapeutic potential and biological role of endogenous antioxidant enzymes in multiple sclerosis pathology. Brain Res Rev 56:322-30.

Schreiner B, Heppner FL, Becher B (2009): Modeling multiple sclerosis in laboratory animals. Semin Immunopathol 31:479-95.

Sorensen PS, Lycke J, Eralinna JP, Edland A, Wu X, Frederiksen JL, Oturai A, Malmestrom C, Stenager E, Sellebjerg F, Sondergaard HB (2011): Simvastatin as add-on therapy to interferon beta-1a for relapsing-remitting multiple sclerosis (SIMCOMBIN study): a placebo-controlled randomised phase 4 trial. Lancet Neurol 10:691-701.

Stanislaus R, Pahan K, Singh AK, Singh I (1999): Amelioration of experimental allergic encephalomyelitis in Lewis rats by lovastatin. Neurosci Lett 269:71-4.

Sternberg Z, Chadha K, Lieberman A, Hojnacki D, Drake A, Zamboni P, Rocco P, Grazioli E, Weinstock-Guttman B, Munschauer F (2008): Quercetin and interferon-beta modulate immune response(s) in peripheral blood mononuclear cells isolated from multiple sclerosis patients. J Neuroimmunol 205:142-7.

Su KG, Banker G, Bourdette D, Forte M (2009): Axonal degeneration in multiple sclerosis: the mitochondrial hypothesis. Curr Neurol Neurosci Rep 9:411-7.

Sueoka N, Suganuma M, Sueoka E, Okabe S, Matsuyama S, Imai K, Nakachi K, Fujiki H (2001): A new function of green tea: prevention of lifestyle-related diseases. Ann N Y Acad Sci 928:274-80.

Taupin P (2008): Adult neurogenesis, neuroinflammation and therapeutic potential of adult neural stem cells. Int J Med Sci 5:127-32.

Theoharides TC (2009): Luteolin as a therapeutic option for multiple sclerosis. J Neuroinflammation 6:29.

Tirosh O, Sen CK, Roy S, Kobayashi MS, Packer L (1999): Neuroprotective effects of alpha-lipoic acid and its positively charged amide analogue. Free Radic Biol Med 26:1418-26.

Trapp BD, Peterson J, Ransohoff RM, Rudick R, Mork S, Bo L (1998): Axonal transection in the lesions of multiple sclerosis. N Engl J Med 338:278-85.

Tron K, Novosyadlyy R, Dudas J, Samoylenko A, Kietzmann T, Ramadori G (2005): Upregulation of heme oxygenase-1 gene by turpentine oil-induced localized inflammation: involvement of interleukin-6. Lab Invest 85:376-87.

Tzima S, Victoratos P, Kranidioti K, Alexiou M, Kollias G (2009): Myeloid heme oxygenase-1 regulates innate immunity and autoimmunity by modulating IFN-beta production. J Exp Med 206:1167-79.

van der Goes A, Brouwer J, Hoekstra K, Roos D, van den Berg TK, Dijkstra CD (1998): Reactive oxygen species are required for the phagocytosis of myelin by macrophages. J Neuroimmunol 92:67-75.

van der Valk P, De Groot CJ (2000): Staging of multiple sclerosis (MS) lesions: pathology of the time frame of MS. Neuropathol Appl Neurobiol 26:2-10.

van Horssen J, Drexhage JA, Flor T, Gerritsen W, van der Valk P, de Vries HE (2010): Nrf2 and DJ1 are consistently upregulated in inflammatory multiple sclerosis lesions. Free Radic Biol Med 49:1283-9.

van Horssen J, Schreibelt G, Drexhage J, Hazes T, Dijkstra CD, van der Valk P, de Vries HE (2008): Severe oxidative damage in multiple sclerosis lesions coincides with enhanced antioxidant enzyme expression. Free Radic Biol Med 45:1729-37.

van Horssen J, Witte ME, Schreibelt G, de Vries HE (2011): Radical changes in multiple sclerosis pathogenesis. Biochim Biophys Acta 1812:141-50.

van Meeteren ME, Teunissen CE, Dijkstra CD, van Tol EA (2005): Antioxidants and polyunsaturated fatty acids in multiple sclerosis. Eur J Clin Nutr 59:1347-61.

Venugopal R, Jaiswal AK (1996): Nrf1 and Nrf2 positively and c-Fos and Fra1 negatively regulate the human antioxidant response element-mediated expression of NAD(P)H:quinone oxidoreductase1 gene. Proc Natl Acad Sci U S A 93:14960-5.

Verbeek R, van Tol EA, van Noort JM (2005): Oral flavonoids delay recovery from experimental autoimmune encephalomyelitis in SJL mice. Biochem Pharmacol 70:220-8.

Weber MS, Zamvil SS (2008): Statins and demyelination. Curr Top Microbiol Immunol 318:313-24.

Witte ME, Bo L, Rodenburg RJ, Belien JA, Musters R, Hazes T, Wintjes LT, Smeitink JA, Geurts JJ, De Vries HE, van der Valk P, van Horssen J (2009): Enhanced number and activity of mitochondria in multiple sclerosis lesions. J Pathol 219:193-204.

Wu CC, Lu KC, Chen JS, Hsieh HY, Lin SH, Chu P, Wang JY, Sytwu HK, Lin YF (2008): HO-1 induction ameliorates experimental murine membranous nephropathy: anti-oxidative, anti-apoptotic and immunomodulatory effects. Nephrol Dial Transplant 23:3082-90.

Youssef S, Stuve O, Patarroyo JC, Ruiz PJ, Radosevich JL, Hur EM, Bravo M, Mitchell DJ, Sobel RA, Steinman L, Zamvil SS (2002): The HMG-CoA reductase inhibitor, atorvastatin, promotes a Th2 bias and reverses paralysis in central nervous system autoimmune disease. Nature 420:78-84.

Yuan R, Maeda Y, Li W, Lu W, Cook S, Dowling P (2008): Erythropoietin: a potent inducer of peripheral immuno/inflammatory modulation in autoimmune EAE. PLoS ONE 3:e1924.

Zenclussen ML, Anegon I, Bertoja AZ, Chauveau C, Vogt K, Gerlof K, Sollwedel A, Volk HD, Ritter T, Zenclussen AC (2006): Over-expression of heme oxygenase-1 by adenoviral gene transfer improves pregnancy outcome in a murine model of abortion. J Reprod Immunol 69:35-52.

Zhang J, Li Y, Cui Y, Chen J, Lu M, Elias SB, Chopp M (2005): Erythropoietin treatment improves neurological functional recovery in EAE mice. Brain Res 1034:34-9.

Zwerina J, Tzima S, Hayer S, Redlich K, Hoffmann O, Hanslik-Schnabel B, Smolen JS, Kollias G, Schett G (2005): Heme oxygenase 1 (HO-1) regulates osteoclastogenesis and bone resorption. Faseb J 19:2011-3.